Biotech's DICTIONARY *Of* Botany

Compiled & Edited by

Meena Francis

BIOTECH

2010

BIOTECH BOOKS

DELHI - 110035

Biotech's

DICTIONARY *Of* Botany

Compiled & Edited by
Meena Francis

First Indian Edition 2005
Second Impression 2010

ISBN 81-7622-130-9

Published by : BIOTECH BOOKS
1123/74, Tri Nagar,
Delhi - 110 035
Phone: 27382765
e-mail: biotechbooks@yahoo.co.in

Showroom : 4762-63/23, Ansari Road, Darya Ganj,
New Delhi - 110 002
Phone: 23245578, 23244987

Printed at : Chawla Offset Printers
New Delhi - 110 052

PRINTED IN INDIA

PREFACE

Love of nature finds its scientific and theoretical expression in the study of 'Botany'. A subject dealing with plants, from unicellular varieties to most complex organisations, botany employs a wide array of concepts and vocabulary. Often, its scientific (read Latin) vocabulary acts as a deterrent to its study. In this dictionary, we have tried to take care of this side of the problem.

Students often wonder why the biological taxonomy is so difficult to remember. The reason is that they are artificially created names that are constructed to give taxonomical information, i.e. the species and the genus names. They are also very frequently a tribute to the discoverer of that species of plant. And, all this information is packed in a few Latin words because Latin used to be the language of the educated classes throughout Europe till the 18th century and is still considered more universal and superior to other languages, by the biologists.

Plants evolved on earth earlier than animals. Not only are animals dependent on plants for food, but their evolution was also made possible by a variety of blue-green algae that cleaned the poisonous atmosphere of the early years.

Structure and physiology of plants is simple as compared to animals, but their cells, which are the building blocks of life, are full or complexity. In this dictionary, we have given short explanations and meanings of various words which are useful to understand the cell and the metabolic processes going on in it.

In the recent years, the study of ecology and genetic engineering has rekindled interest in botany. The interdependence of human and natural community was never so acutely felt, but genetic engineering has opened immense potentialities. Welcome to the '**brave new world.**'

- **a / an**
  a prefix indicating the absence of a feature, as for example in alete.

- **ABA (abscisic acid)**
  a plant growth regulator found in plants that acts as a signaling molecule which helps to control overall growth and physiology. It regulates responses to environmental stresses such as drought, cold, salinity and heat.

- **abaxial**
  situated or directed away from an axis or away from main axis, often underside of branch.

- **abc soil**
  soil that has a surface layer, subsoil and a lower layer of loose rock material. The layers are said to be its A, B and C-horizons.

- **abciss**
  detach, separate from main structure.

- **aberrant**
  this describes a plant or species that is different in some way from the group in which it is placed.

- **ablastous**
  means without germ or bud.

- **abnormal**
  different in some manner from other specimens in the taxon.

- **abortive**
  development arrested while incomplete or imperfect.

- **abporal lacuna / abporal lacunae**
  a lacuna in a lophate pollen grain situated at the end of an ectoaperture that is divided by sexinous ridges into two or more lacunae. Example Sonchus oleraceus (Composite).

- **abruptly pinnate**
  this describes a pinnate leaf that ends without an odd leaflet or tendril. It ends in a matched pair of leaflets.

- **abscise**
  to cut off. Hence, abscission, the normal separation of a leaf, fruit or flower from a plant.

- **abscission zone**
  a region of the petiole.

- **absorption**
  the process of taking up water through the roots.

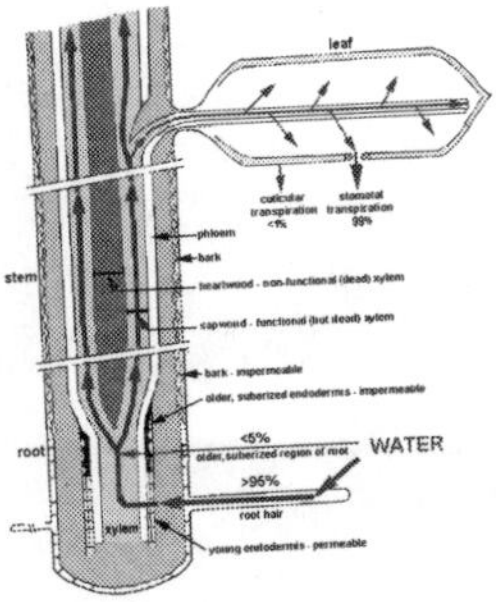

- **acalymmate**
describing tetrads or polyads in which the sexine/ectexine of each monad is well differentiated, but does not form a single continuous envelope around the unit. Example Drosera (Droseraceae).

- **acantha**
a botanical word that refers to a spine, thorn or prickle.

- **acanthocarpa**
bearing spiny fruit. Also echinocarpa.

- **acanthodes**
full of thorns.

- **acanthomamilla**
a biform sculptural element consisting of a hemispheroidal base, surmounted by a sharply contracted spine. Examples Acinosporites, Diblolisporites.

- **acarpous**
this describes a plant that is sterile.

- **acaulescent**
1. without a visible stem.
2. a plant that is stemless. The word Acauline is also used.
3. (Gr. *a*, without; kaulos, stalk) having no stem or seemingly without a stem.

- **accessory**
subsidiary, like a non-chlorophyll pigment.

- **accessory flower parts**
sepal and petal organs found on flowers. The sepals and petals are not essential for pollination but may aid in attracting insects or other organisms.

- **accretive**
increasing or adding by growth. Continued development within.

- **accumbent**
1. of a pair of cotyledons curving back towards the hypocotyl so that their edges are turned towards the hypocotyl.

2. lying against. Mainly used to describe the first sprouts of an embryo when they lie against the body of the seed.

- **aceae**

new Latin suffix, meaning 'the plants of the nature of', which is used to form names of families of plants. Examples are Liliaceae, Rosaceae etc.

- **acephate**

a man-made, systemic insecticide that is used to control numerous pests such as aphids, scale and thrips, of most ornamental plants.

- **acerose**

needle-shaped; narrower than subulate, tapering more gradually to a point.

- **acervate**

growing in heaps or in compact clusters.

- **acetabuliform**

concave, with slightly inturned border.

- **acetolysis**

a widely used technique for preparing pollen and spore exines for study.

- **achene**

this is a small, dry, hard, one-celled, one-seeded fruit that stays closed at maturity.
(Gr. *a*, not *chainein*, to gape) any small, dry fruit with one seed whose outer covering (pericarp) does not burst when ripe.

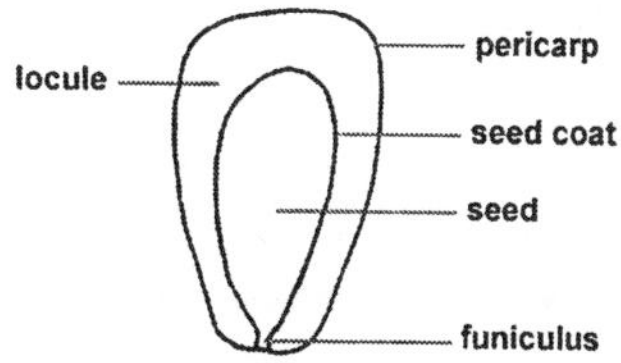

Achene (e.g. fruit of a sunflower)

- **acicular**

needle shaped, applied to an elongated tapering structure with a circular of elliptical cross section. Or needle-shaped.

- **acid soil**

soil with a pH value of less than 7 or lacking in lime. See **alkaline and neutral.**

- **acidify**

1. to lower the soil's pH to meet the level of acid soil.
2. the level of acid present in the soil.

- **acinaceous**

consisting of or full of kernels.

- **acinaciform**

shaped like a scimilar (usually describes a leaf shape). An acinaciform leaf has one edge convex and thin and the other edge straight and thick. This term is also used to describe the shape of the pods of some beans.

- **acinarious**
covered with little round, stalked, grape-like sacs, as in some algae.

- **aciniform**
having the form of grapes or growing in grape-like clusters-acinose.

- **acolpate**
without colpi.

- **acritarch**
1. small microfossil of unknown or varied biological affinity, consisting of a central cavity enclosed by a wall of single or multiple layers.
2. an informal utilitarian, 'catch-all' category without status as a class order or other suprageneric uni, comprised of 'small microfossils of unknown and probably varied biological affinities consisting of a central cavity enclosed by a wall of single or multiple layers and a chiefly organic composition'.

- **acrocarpous**
used to describe an erect habit in mosses.

- **acrogen**
a plant that increases by growth at the summit or terminal buds only. The acrogens form a division of the Cryptogams, including ferns, fern allies, liverworts and mosses.

- **acrolamella / acrolamellae / acrolamellate**
a leaf-like tapering segment on the proximal pole of a megaspore. Example Arcellites. An acrolamella is a special form of gula, represented by leaf-like segments.

- **acropetal**
1. from the base to the apex.
2. (Gr. *akros*, summit L. *petere*, to seek) developing upward from the base toward the apex.

- **actinomorphic**
(Gr. *aktis*, ray *morphe*, form) descriptive of a flower or set of flower parts which can be cut through the center into equal and similar parts along two or more planes. Having radial symmetry.

- **actinostele**

1. a type of protostele that looks like a star in cross section.
2. a protostele having a xylem core in the form of radiating ribs, as viewed in transverse section.

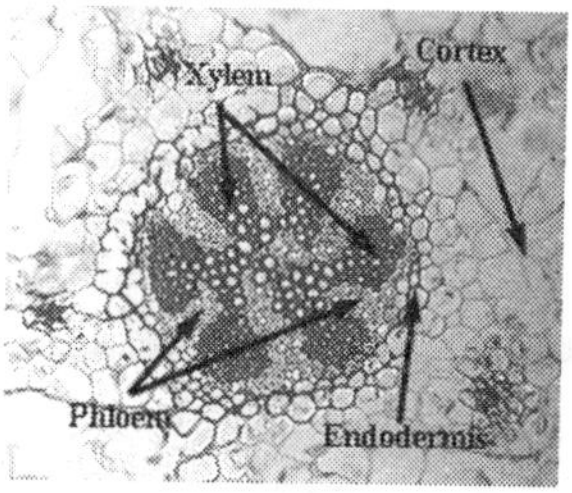

- **activator**

any substances, such as microorganisms and nutrients, which are added to a compost pile to speed the breakdown of organic matter. Also called compost activator and compost inoculant.

- **active trap**

a trap which is capable of some degree of movement to catch or detain its prey. See **passive trap.**

- **actuopalynology**

1. the study of extant palynomorphs which are either living, still retain their cell contents or whose cell contents have been removed by maceration. Includes mellisopalynology, pollination ecology, aeroallergy and criminology. Exclusive of paleopalynology.
2. the study of pollen grains and spores of extant plants.

- **aculeate**

armed with prickles or bearing a sharp point.

- **aculeus**

(L. *aculeus*, prickle) a prickle growing from bark. pl. aculei.

- **acumen**

(L. *acumen*, a point, a sting) the point of an acuminate leaf.

- **acuminate**

1. with a long pointed structure formed by abrupt narrowing at an apex.
2. tapers toward the end into a long point.
3. a drawn out into a long point, tapering point.

- **acuminose**

rather less gradual than acuminate.

- **acute**

ending in a sharp point, forming an angle of less than 90 degrees. Sharp-pointed.

- **adaptive radiation**
  the diversification of a species under different, local environmental pressures to which it becomes adapted, eventually leading to the emergence of new species.
- **adaxial**
  on the upper surface. Situated or directed toward an axis.
- **additive**
  something that is added to a fertiliser to improve its chemical or physical condition.
- **adelphous**
  means having stamens united into sets. Used mostly in combination, as in monoadelphous.
- **adhesion**
  of two dissmiilar parts touching each other but easily separated and not fused or grown together.
- **adnate**
  in botany, grown together or attached.
- **adnation**
  fusion of unlike parts such as the *labellum* (of an Orchid) with the *column*.
- **adnexed**
  in botany, annexed or touching, but not attached.
- **adobe soil**
  dense, clay-like soil, also called gumbo, that is also high-alkaline.
- **adosculation**
  in botany, the fertilisation of plants by the falling of the pollen on the pistils.
- **adpressed**
  lying close and flat against.
- **adsorb**
  to accumulate on the soil's surface.
- **adsortion**
  condensation on the soil's surface.
- **adventitious buds**
  buds appearing in an unusual place. Example, buds on leaves.
- **adventive**
  a plant that has been introduced, but isn't naturalised yet.
- **aerate**
  to expose to the air or to charge with air.
- **aeration**
  an exchange of air in the soil with air from the atmosphere. Earthworms do this as does digging and turning the soil to loosen it.
- **aerator**
  any implement that is used to break up compacted soil and facilitate air and gas exchange.

- **aerator sandals**
  sandal-type shoes or spikes strapped onto regular shoes used to break up compacted soil in lawns.

- **aerial root**
  1. arising above ground, often from an axil. Provides a means for the plant to absorb moisture from the air.
  2. this is a root that grows from the stems or shoots of a plant and helps to sustain it by absorbing moisture from the air. Many Orchids and other tropical plants have aerial roots.
  3. a root that grows in air, common in epiphytes.

- **aerial stem**
  an erect stem arising from a horizontal rhizome.

- **aeroallergy**
  the branch of medicine concerned with the seasonal occurrence, abundance and allerogenic effect of spores and pollen.

- **aerobic**
  living or acting only in the presence of air.

- **aeropalynology**
  the study of palynomorphs found in the atmosphere.

- **aestivation / estivation**
  spending the summer months in a dormant state.

- **affixed**
  of aquatic plants, rooted in soil but growing in open water.

- **aggregata**
  fine soil particles that are held together in a single mass.

- **air layering**
  this is a method of propagation in which roots are induced to form around a plant's stem. This is done by removing a very narrow strip of bark from around the branch or stem, about 8 to 12 inches from the tip of the shoot or at that same distance from the tip, a cut can be made about a third way through the branch in an upward direction and a sliver of wood can be inserted into the cut to keep it open. Next, a bundle of moist sphagnum moss is tied securely around the cut area. The moss must always remain moist (this may be done by wrapping it in polyethylene film) and the roots of the plant somewhat dry. New roots will soon sprout from the incision. The new plant is then cut off below the moss, potted and kept in a humid atmosphere until it is established.

- **air plant**
  an *epiphyte*. Many epiphytic Orchids and Bromeliads are

popularly called air plants.

- **air/water leaves**
in aquatic plants, leaves that grow below the water surface are distinct from those formed above the surface.

- **akaryotic**
without a true nucleus.

- **akinete**
enlarged cell with food reserves and thick cell wall which may undergo dormancy, e.g. in Cyanobacteria.

- **albuminous cells**
a cell the phloem that is connected to a sieve cell by numerous plasmodesmata.

- **aldrovanda**
a montypic genus of aquatic plant having 'sprung' traps similar in form to those of Dionaea .

- **alete**
describing a spore without a laesura.

- **algae**
simple plants in the Division Protista, consisting of single cells to multicellular organisms, arranged in colonies or as filaments and containing photosynthetic pigments. Many common algae are green because their cells contain chlorophyll.

- **alginate**
salt form of alginic acid, a polysaccharide colloid produced in walls of Phaeophyceae made of mannuronic acid and guluronic acid units.

- **alkali soil**
highly alkaline soil with a pH of 8.5 or higher. Soluble salt or a mixture of soluble salts present in a soil.

- **alkaline soil**
soil with a pH value more than 7 or lime-rich.

- **alkaloid**
complex cyclic organic plant metabolites, usually containing nitrogen, with a bitter taste and marked physiological activity on animals. Many of the 7000 known alkaloids are basic (alkaline). Their function

may be to protect the plant from non-adapted herbivores, either by virtue of their unpleasant taste or their toxicity. Examples include atropine, caffeine, cocaine, colchicine, mescaline, morphine, nicotine, quinine, strychnine.

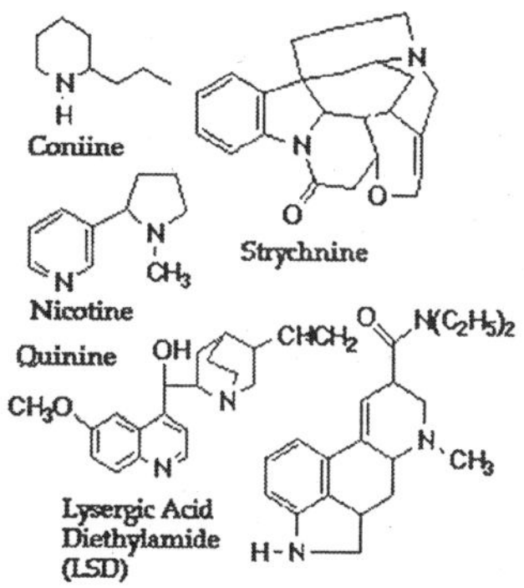

- **allelochemicals**
  compounds that have an allelopathic effect.

- **allelopathy**
  (Gr. *allelon*, one another *pathos*, suffering) the influence or effect of one living plant upon another; refers to biochemical interaction between all types of plants and its effect depends on a chemical compound being added to the environment.

- **allochtonous**
  something that has been transported.

- **allogamy**
  the process of cross-fertilisation in which the pollen of one plant fertilises the flower of another.

- **alloparasite**
  parasite not related to its host

- **alluvial soil**
  referring to the sediment deposited by flowing water.

- **alluvium**
  clay, silt, sand and/or gravel deposited on land by running water.

- **aluminum silicates**
  compounds that contain aluminum, silicon and oxygen.

- **amino acid**
  a nitrogen containing organic compound that is linked together to form a protein molecule.

- **ammonia**
  1. a colourless gas made up of one part nitrogen and three parts hydrogen.
  2. a gas that can be liquified under pressure to be used as a fertiliser.

- **ammonification**
  the act of forming ammonium compounds from organic material by various organisms.

- **ammonium ion**
  a form that nitrogen takes in many commercial fertilisers.

- **ammonium nitrate**
  1. an organic fertiliser that will add nitrogen to the soil.
  2. the form that nitrogen takes in numerous commercial fertilisers.

- **ammonium sulphate**
  an inorganic fertiliser applied to soil (20.6 % N.).

- **amoeba**
  shells or tests enclosing the protoplasm characterise protozoans in the subclass Testacealobios of the Rhizopoda. The tests are 5 - 250 micrometres in size and are generally resistant to routine pollen extraction. The organisms are abundant in mesic habitats ranging from damp soil to lakes. The tests are urn-shaped to flattened, with various sculpturing and ornamentation.

- **amorphous**
  lacking definite structure or shape.

- **amphibious**
  able to grow in both water and on land or growing with part of the plant in water or mud and part in the air.

- **amphicarpous**
  refers to a plant that produces two classes of fruit that differ either in form or in time of ripening.

- **amphiphloic**
  having phloem on both sides of the xylem.

- **amphistomatic**
  stomata in upper and lower epidermis.

- **amphitropous**
  having the seed embryo curved or coiled back along the supporting stalk so that its base and the opening of the pollen tube bend close to each other.

- **amplexicaul**
  1. nearly surrounding or embracing the stem, as the base of some leaves.
  2. clasping the stem, eg. sessile leaves that connect broadly around the stem.

- **amplexifoliate**
  having leaves that clasp the stem.

- **ampulla**
  a small bladder of flask-shaped organ attached to the roots or immersed leaves of some aquatic plants, as in Nepenthes or Utricularia.

- **amyloplast**
  1. contains large amounts of starch, but no chlorophyll, specialised for storage.
  2. colourless organelle related to starch production.

- **ana-**
  a prefix indicating the position of features, such as apertures, on the distal face. Examples Nymphaea (Nymphaeaceae), Sparganium (Typhaceae).

- **anadromous**
  having the first lobe or segment of a pinna arising basiscopically in compound leaves.

- **anaerobic**
  having very little or no oxygen levels. Some aquatic plants can grow anaerobically while producing new growth in water.

- **anaerobiosis**
  (L. Gr. *an*, without; *aer*, air *biosis*, manner of life) life in the absence of air or free oxygen.

- **analogous, analogy**
  opposite to homologous, correspondence in function between anatomical parts of different structure and origin, analogous, exhibit analogy (synonym similar).

- **anastomose**
  to join together e.g. veins.

- **anastomosis**
  (Gr. *ana*, up to *stoma*, mouth) connecting by cross-veins and forming a network.

- **anatomy**
  (L. *anatomia*, dissection) the branch of morphology that deals with the structure of plants, esp. the internal structure as revealed by the microscope.

- **anatropous**
  turned back on itself.

- **anazonasulculate**
  describing a pollen grain with a ring-like sulculus situated between the equator and the distal pole.

- **anceps**
  two-edged.

- **ancyrate**
  bearing sub-cylindrical or tapering processes which divide at their distal extremities into anchor-shaped or multifurcate tips.

- **androdioecious**

having hermaphrodite (bisexual) flowers on one plant of a species and male flowers on another plant of the same species, but no corresponding form with only female flowers.

- **androecium**

1. the male organs of a flower; the assemblage of stamens.
2. (Gr. *andros*, man *oikos*, house) male reproductive organs of a plant. A collective term applied to all structures of the stamen whorl or whorls.

- **androgynal**

(Gr. *andros*, man *gonos*, woman) bearing staminate and pistillate flowers on the same parent stem.

- **androgynous**

1. staminate flowers above the pistillate flowers in the same inflorescence.
2. with stamens and pistils in the same flower or inflorescence.

- **androphore**

1. the stalk or column supporting the stamens of certain flowers usually formed by a union of the filaments, as in the Malvaceae and in many genera of Leguminosae.
2. (Gr. *aner*, man *phoros*, carrying) a support or column, formed by fusion of filaments, on which the stamens are borne.

- **anemone centered (of flowers)**

flowers or flower heads in which the central petals or flowers (modified *stamens*) form a pincushion-like mound and the outer rim or petals or ray florets are flat and spreading, as in some chrysanthemums.

- **anemophilous**

wind pollinated; refers to flowers that depend on the wind for pollination. Anemophilous flowers are ordinarily unnoticeably small, uncoloured and don't secrete honey; however, they produce an abundance of pollen. Examples include the flowers of grasses, pines, sedges etc.

- **angiocarpian**

a plant that has its fruit enclosed in a husk.

- **angiocarpous**

having a fruit enclosed within a distinct covering, as the filbert within its husk.

- **angiosperm**

1. any flowering plant that reproduces by seeds enclosed in a protective seed vessel. The term angiosperms is given to the larger of the two divisions of seed-bearing plants, in contrast to gymnosperms, which lack flowers and in which the seeds are naked (Examples are Coniferae, Cycadaceae etc.).
2. a flowering plant with ovules (seeds) enclosed in an ovary (fruit). Angiosperms developed about 100 million years ago (late Jurassic) and have become the dominant type of terrestrial plant, represented by at least 250,000 species, including both monocotyledons (eg. Agave) and dicotyledons (eg. Cactaceae).

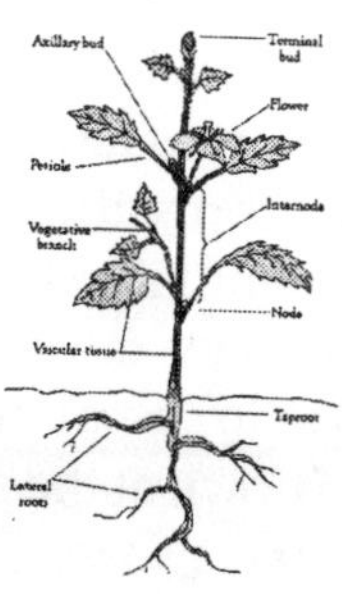

3. (Gr. *anggeion*, vessel *sperma*, seed) a major division of the plant kingdom, commonly called flowering plants as their reproductive organs are in flowers, having seeds which develop in a closed ovary made of carpels, a very reduced gametophyte and endosperm develop from a triple fusion nucleus. pl. Angiosperms.

- **angiospermous**

having seeds enclosed in a seed vessel, for example, this refers to the poppy, the rose and most flowering plants, opposed to gymnospermousee or naked-seeded.

- **angulaperturate**

describing an equatorially aperturate pollen grain with the apertures situated at the angles of the outline in polar view. Example Corylus (Betulaceae).

- **angular**

having projecting longitudinal angles, undulate, but sharply angled rather than gently rounded.

- **angular collenchyma**

cell wall is thickest in the corners.

- **angustata**

with a narrow stem. Attenuate, tapering smoothly and concavely to a point.

- **angustimurate**
with narrow muri. Erdtman restricted the term to muri up to 1/5th of the diameter of the lumina.

- **anhydrous**
dry or lacking water.

- **anisodiametric tetrad**
general term for tetrads which members are differing in size. Example seed-megaspore tetrads.

- **anisogametes**
motile gametes which differ slightly in size.

- **anisogamy**
reproduction by motile gametes that differ in morphology or behaviour.

- **anisophyllous**
having the leaves of a pair of two different shapes and sizes, as in conifers and many aquatic plants.

- **anisopolar**
synonym of heteropolar.

- **annual**
1. a plant that naturally completes its life cycle, from seed to seed, in one year.
2. a plant that completes an entire life cycle in one growing season. A life cycle includes germination, growth to maturity, blooming, producing seeds and death.
3. (L. *annualis*, yearly, from *annus*, year) a plant which completes its life history within a year.

- **annual ring**
1. in trees and other woody plants, this is the layer of wood produced each year that can be seed when the wood is cut into a cross section. The number of rings equals the age of the tree.
2. the portion of a woody stem produced in one year.
3. layers of wood laid down each year by the vascular cambium of woody plants, typically in trees. During the season of active growth in the vascular cambium, xylem cells are being added to the circumference of the xylem tissue. At the start of the growing season the cells grow fast and are large, but as the season progresses towards the dormant period fewer smaller cells are produced and growth eventually stops as the plant becomes dormant. The difference in size of the xylem cells laid down at the start and later in the growing season

forms a line within the wood marking the boundary of an annual ring. The size of the xylem cells is also affected by the growing conditions. Hence changes in the width of annual rings can indicate changes in the climate.

- **annulate**

1. ringed or banded.
2. specialised of some fern sporangia that are involved in the opening of the sporangium. On drying out the cells of the annulus contract and the sporangium ruptures, releasing the spores.
3. a ring-like part, structure or membrane.
4. the elastic ring that surrounds the sporecase of most ferns.
5. the raised ring around the center of a succulent flower.
6. thick-walled ring of cells on the sporangium.
7. an area of the exine surrounding a pore that is sharply differentiated from the remainder of the exine, either in ornamentation or thickness.

- **anomalous secondary growth**

secondary tissue that are not produced by a vascular cambium.

- **anoxia**

(L. *an*, not and oxygen) lack of oxygen or not enough oxygen.

- **anterior**

the underside of a leaf; otherwise called inferior or lower; the opposite of the posterior. Also used in flowers, as on a Snapdragon, where the lower lip is anterior and the upper lip posterior.

- **anteturma / anteturmae**

an artificial grouping of fossil spores and pollen in the turma-system of Potonié.

- **anther**

1. the male part of a flower producing pollen, usually on a filament. Anther and filament together are described as a stamen.

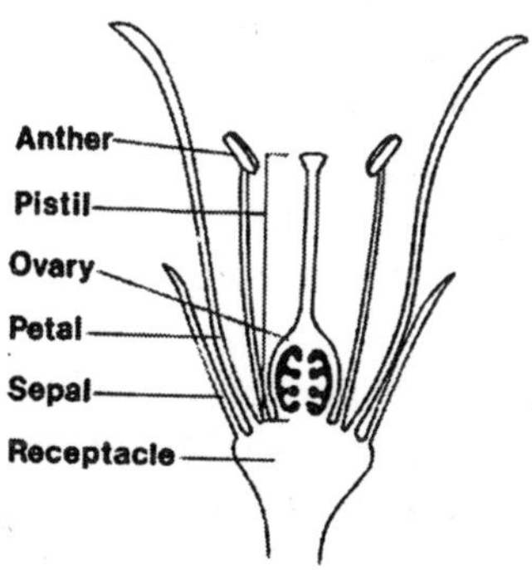

2. the part of a *stamen* that produces pollen, usually located at the end of a thin stalk in the center of a flower.

3. pollen-bearing portion of the stamen
4. (Gr. *antheros*, flowery, from *anthein*, to bloom) the top of the stamen, usually elevated by means of a filament, which contains the pollen.
5. the apical portion of the stamen containing pollen.

- **antheridia**

reproductive structure that produces male gametes (sperm).

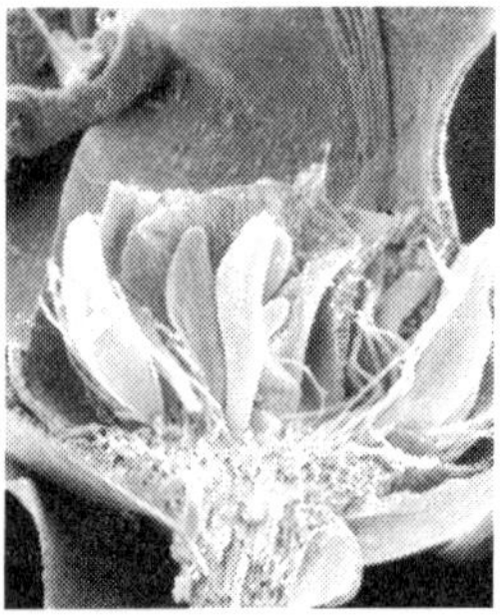

- **antheridium**

1. pl. antheridia- the structure on the pteridophyte (fern) prothallus that produces the mobile male gametes.
2. the male reproductive organ in spore-bearing plants such as ferns, mosses, algae and fungi. Corresponds to the anther in seed-bearing plants.
3. the male sex organ producing the sperm.
4. cell that produces motile male gamete.

- **antherozoid**

the thread-like substance produced by the male with which the female organs are fertilised.

- **anthesis**

1. the action of opening a flower or period of time during which a flower is open.
2. the period during which a flower opens or the act of a flower opening; coming to full bloom.
3. (Gr. *anthesis*, bloom, from *anthein*, to bloom) stage or period during which the flower bud is fully open, flowering.

- **anthocyanin**

1. pink, purple, violet or blue water-soluble pigments colouring leaves and other parts. These pigments are sensitive to acidity or alkalinity (pH) turning red in acid and blue in alkali.
2. the dissolved colouring matter (glycoside pigments) in blue flowers. Also anthocyan.

- **anthocyanins**

red, blue or purple pigments that are water soluble.

- **anthography**

the description of flowers.

- **antholeucin**
  the dissolved colouring matter in white flowers.

- **anthophore**
  an elongated segment as the base of a flower, between the stem and the base of the petals, stamens and pistil, as in pinks (Dianthus) or catch-fly (Silene). Also called anthophorum.

- **anthoropic soil**
  soil worked by humans in a way to produce new characteristics that make it different from the original natural soil.

- **anthoxanthin**
  the yellow or orange colouring matter of yellow flowers and fruit.

- **antibiosis**
  not alive or The absence of life.

- **antibiotic**
  a substance that destroys or inhibits the growth of bacteria and other microorganisms.

- **anticlinal**
  perpendicular to the surface.

- **antipetalous**
  1. having stamens opposite the petals.
  2. standing opposite to sepals, applied to stamens.

- **anulus**
  orthographical variant of annulus.

- **apatite**
  the main mineral found in phosphate rock. An inorganic compound found in bones. A component of the fertiliser, bone meal.

- **aperture/aperturate**
  a specialised region of the (sporoderm, that is thinner than the remainder of the sporoderm and generally differs in ornamentation and/or in structure.

- **aperture membrane**
  the exine which forms the floor of an ectoaperture. Depending on the type of ectoaperture these may be colpus membranes or pore membranes, usually representing the nexine. The membrane may be partially or completely covered by an operculum.

- **apertures**
  any thinning or modification of the wall of spores or pollen that serves as an exit for its contents. In embryophyte spores these are the monolete or trilete laesura. In pollen these are the furrows (colpi) and pores. The

arrangement and number of pores and furrows are chief criteria for identifying pollen classes.

- **apetalous**

1. without petals or corolla.
2. (Gr. *a*, without *petalon*, leaf) having flowers without petals, having no corolla.

- **apex**

1. the top part of a plant or structure, the growing point.
2. the end of a leaf or growing tip of a shoot.
3. the tip of something, usually a shoot, root or leaf.
4. tip, pointed end.
5. a general term for the tip of an organ. In fossil spores applied to the tip or corner of a trilete spore.

- **apheliotropic**

turning away from the light, applied to shoots or other parts of plants, as in roots that turn away from the sun, opposite heliotropic.

- **aphyllous**

without leaves. Applied to flowering plants that are naturally leafless, such as many kinds of cactus.

- **apical**

1. located at the tip or highest point.
2. at or near the apex or top.
3. (L. *apex*, the tip or top of a thing) at the tip or summit.
4. of the apex.

- **apical cell**

the single initial in the apical meristems of most seedless vascular plants.

- **apical dome**

the part of the apical meristem interior to the leaf primordia.

- **apical dominance**

inhibition of lateral bud development along a shoot by the terminal bud or growing point.

- **apical initial**

a cell that, when it divides, leaves one of the daughter cells in the apical meristem.

- **apical meristem**

1. the un-differentiated tissue at the tip of a shoot or root, that by division produces the precursors of the differentiated tissues, including vegetative, reproductive or floral organs.
2. embryonic, totipotent tissue in the tips of the roots and shoots of plants.

- **apicalfeld**

synonym of apocolpial field.

- **apiculate**

1. (LL. *apiculatus*, point) terminated abruptly by a small, distinct point, an apiculus or apicule.

2. describing the shape of a pollen grain that has slightly protruding polar caps. Example Adonis aestivalis (Ranunculaceae).

- **apiculate elements**
  projections from the general surface. Examples bacula, pila, verrucae, spines.

- **apicule**
  a short, sharp point.

- **apocarpous**
  1. an ovary of a fruit composed of one or more simple and distinct pistils, as in Buttercups.
  2. (Gr. *apo*, away *karpos*, fruit) having separate carpels.

- **apocolpial field**
  a region at the pole of a parasyncolpate pollen grain, delimited by the margins of the anastomosing colpi.

- **apocolpium / apocolpia**
  a region at the pole of a zonocolpate pollen grain delimited by lines connecting the apices of the colpi.

- **apocolpium index**
  the ratio of the distance between the apices of two ectocolpi (d) of a zonocolpate pollen grain to its equatorial diameter (D). Synonym of polar area index.

- **apogamous**
  the nature of or characterised by apogamy.

- **apogamy**
  1. generally, the absence of sexual reproductive power, the plant perpetuating itself only by vegetative means, as buds, bulbs etc.
  2. producing sporophytes from a gametophyte without fertilisation.
  3. development of an organism without gamete fusion (fertilisation).

- **apolar**
  describing pollen and spores without distinct polarity. Example Plantago (Plantaginaceae).

- **apomeiosis**
  nuclear division without meiosis.

- **apomixis**
  reproduction involving generative tissues but not dependent on fertilisation. (Gr. *apo*, away; *mixis*, a mixing) in general, reproducing without sexual reproduction; often used to denote seed production without a sexual process having been involved.

- **apomorphy**
  same as derived trait.

- **apophyllous**
  having distinct leaves, applied to a whorled flower with distinct sepals and petals, opposed to gamophyllous.

- **apophysis**
  an offshoot or projecting part.

- **apoporium / apoporia**
  an area at the pole of a zonoporate pollen grain that is delimited by a line connecting the borders of the pores. The term contrasts with apocolpium and is most meaningful in pollen grains with large pores.

- **apospory**
  producing gametophytes directly from a sporophyte without producing spores.

- **appendage**
  1. that which is attached to something as a proper part of it; a subordinate, attached part of anything. Specifically, any subsidiary part added to another part, as hairs and glands to a stem or leaf.
  2. a projecting part of a structure.

- **applanate**
  flattened out or horizontally expanded.

- **appressed**
  1. turned upwards and lying closely, flat against something e.g. spines of some cacti.
  2. pressed closely against; fitting closely to, lying flat against.

- **aqua ammonia**
  a water solution of ammonia.

- **aquatic**
  growing in water. Plants that are naturally found growing in water at or above the surface of the soil. There are different types of aquatic plants. Some are emergent or only partially submerged, such as Cattails. Some are completely submerged, such as Anacharis; some float with roots anchored to the soil under water, such as Water Lilies, some float freely with roots free in the water, such as water Hyacinths.

- **aquatic plants**
  plants that must grow in water whether rooted in the mud or floating without anchorage, plants that must complete part or all of their life cycle in or near the water.

- **aquatic vascular plants**
  aquatic plants containing the conductive vascular tissue, phloem and xylem.

- **aquifer**

a water-bearing formation, usually underground, through which water can move easily.

- **arabidopsis**

the first flowering plant to have its entire genome sequenced. Commonly known as thale cress, a relative of the mustard plant, it has become a model organism in plant biology research because Arabidopsis has almost all of the same genes as other flowering plants and has relatively little 'non-coding' DNA.

- **arachnoid**

1. cobweb-like.
2. (Gr. *arachme*, spider, cobweb, *eidos*) like a cobweb, covered with or consisting of soft fibers or hairs so entangled as to give a cobwebby appearance.
3. covered with long hairs like a cobweb.

- **araucariod**

a member of the family of conifers Araucariaceae (or genus *Araucaria*). The monkey puzzle tree is a present day member.

- **arbour**

a bower formed by trees, vines or shrubs intertwined or trained over a latticework trellis, so as

to make a leafy roof. They are usually provided with seats, any shaded walk. Arbours are used to provide support for plants, to create shade or to frame a view.

- **arboraceous**

pertaining to or of the nature of trees.

- **arborescent**

1. tree-like in size and growth habit.
2. resembling a tree, tree-like in growth, size or appearance. Having the nature and habits of a tree. Branching like a tree.

- **arboretum**

a place in which trees and shrubs, especially rare ones, are grown for scientific or other purposes. A botanical tree garden.

- **arboriculture**
  the cultivation of trees. The art of planting, training, pruning and cultivating trees and shrubs for decorative purposes.

- **arborist**
  a specialist in cultivating trees, in diagnosing tree diseases, performing tree surgery and maintaining trees.

- **arbuscle**
  a dwarf tree or tree-like shrub, in size between a tree and a shrub.

- **archaeological palynology**
  the analysis of pollen, spores and other palynomorphs from archeological sites. Applications include the reconstruction of prehistoric diet, funary practices, artifact function and source, archaeological feature use, cultivation and domestication of plants and human impact on vegetation.

- **archegonia**
  reproductive structure that produces female gametes (egg).

- **archegonium**
  1. pl. archegonia- the structure on the pteridophyte (fern) prothallus that produces the sessile female gametes.
  2. the female sex organ producing the egg.

- **arcuate**
  1. in curved like a bow, crescent-shaped.
  2. a L. *arcuatus*, pp. of *arcuare*, to arch, bend like a bow, from arcus, a bow/ bent or curved the form of a bow.

- **arcus / arcus / arcuate**
  a locally thickened band of sexine that extends in a sweeping curve from one aperture to another. Example Alnus (Betulaceae).

- **areola / areolae / areolate**
  feature of ornamentation in which the sexine/ectexine is composed of circular or polygonal areas separated by grooves which form a negative reticulum.

- **areoles**
  the spaces formed by a vein network.

- **aril**
  1. an outer covering or sometimes merely an appendage, of a seed formed from the funicle. May be hard, waxy or pulpy.
  2. applied to the coverings or appendages of seeds.

3. (Fr. *arrile*, Sp. *arillo*, L. *arilli*, dried grapes, from *aridus*, dry) an additional covering that forms on some seeds after fertilisation and developing from the stalk of the ovule.

- **arista**

  an awn or bristle.

- **aristate**

  terminating in an abrupt with an awn or bristle. Bearded, bristle tipped. a (L. *arista*, awn) awned; having an awn.

- **aromatic**

  with a resinous, spicy or distinctive smell.

- **articulate**

  1. jointed or with points of apparent separation, usually marked by a swelling, line or sudden change in colour.
  2. provided with joints or nodes, at which points separation takes place naturally. Jointed, consisting of segments united by joints. An example is Christmas Cactus (Schlumbergera bridgesii)
  3. generally meaning having a joint as in leaves, leaflets or stems, as in heterophyllous species of *Selaginella* or having a swollen area, often discolored, at the point of branching of the stem.
  4. (L. *articulatus*, jointed, pp. of *articulare*, to join) having joints, provided with places where separation may take place.
  5. having clearly distinguishable parts.
  6. jointed or segmented, e.g. Calliarthron or Halimeda.

- **artificial light**

  light other than sunlight, often fluorescent tubes, used to grow plants.

- **ascending**

  growing upwards.

- **ascidium**

  1. any tubular, horn-shaped or pitcher-like formation of a plant, usually arising from the union of the margins of a leaf or other organ. The ascidium, commonly known as a pitcher, is usually covered with a lid and contains a fluid in which insects are drowned and dissolved. Examples are the pitcher plants (Nepenthes and Sarracenia).
  2. the tubular trap formed by a pitcher; the opening of such a trap.

- **asclepius - (Latin *aesculapius*)**
  the ancient Greek mythical god of healing from whom the name of the family *Asclepiadaceae* is derived. The *Asclepiadaceae* have recently been subsumed under *Apocynaceae*.

- **ascomycete**
  a large group of fungi including yeasts, milldews and truffles, with hyphae divided by septa and producing spores sexually within a typically cylindrical membrane forming a spore case or sac (ascus).

- **asepalous**
  (Gr *a* without L. *pelaıum*, petal) without sepals.

- **asexual**
  1. without the involvement of fertilisation, propagation by division or the production of bulbils, offsets or stolons.
  2. not-sexual; having no sex.
  2. 3. produced by other than sexual processes, as asexual reproduction. Asexual reproduction is any process of propagation that isn't effected by means of sexual organs, as when propagation is carried out by buds, offshoots, bulbs etc.

- **asparagus-knife**
  a tool for prying and pulling out long-rooted plants. Also called dandelion weeder or fishtail weeder.

- **asperifoliate**
  having rough-feeling leaves.

- **asperous**
  rough to the touch.

- **aspis / aspides / aspidate**
  an orthographical variant of aspidate prominently protruding thickening of the exine around a pore. Examples Betula (Betulaceae), Dorstenia (Moraceae). An aspis is a special form of annulus, represented by a thickening rather than a thinning.

- **aspis channel**
  synonym of pore canal.

- **assimilation**
  the conversion of a substance taken into the living tissues of plants.

- **astrosclereids**
  star shaped, with several projecting arms.

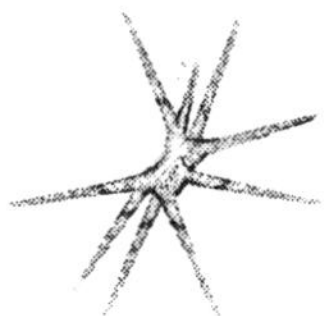

- **atactostele**
  a complex three dimensional network of vascular bundles.

Superficially bundles appear to be scattered.

- **ataxonomic**

an approach in which interpretations are not based on taxonomic assignment; for example if a fossil taxon has the same generic name as an extant taxon, interpretations about the ecology of the fossil taxon are not based on the ecology of the extant taxon.

- **ate**

a suffix for possession of. Example porate.

- **atrium**

1. synonym of fastigium.
2. a space within the aperture of a compound pore that has a much larger endopore than the ectopore, so that the pore canal widens towards the interior of the grain. Example Myrica (Myricaceae).

- **attar**

a general term used for a perfume from flowers. In Europe, it denotes only the attar of roses, an essential oil. It takes 150 pounds of roses to provide less than an ounce of attar. The fragrance is too strong when it is concentrated and must be diluted to smell nice.

- **attenuate**

1. gradually narrowed or tapered.
2. tapering gradually to a narrow point.
3. (L. *attenuare*, to thin) gradually narrowed to a long point at apex or base.
4. narrowing gradually to a point.
5. diminishing in width.

- **a-type tetrad**

a tetrad in which the aborted cells (1-3 in number) are clearly associated with the fertile grain(s). Example Cyathodes juniperina (Epacridaceae).

- **auger**

a tool for boring a hole in the soil. Different-sized augers, powered by hand or electric drills, are used in planting or transplanting seeds, seedlings or bulbs or in fertilising shrubs, trees and ground covers.

- **aurea**

golden.

- **auricle**

ear-shaped projection or appendage. Ear-shaped lobes, often paired, as on the *labellum* or *column* of an Orchid.

- **auricle**

1. (L. *auricle*, small ear) any ear-like lobed appendages.
2. an ear-shaped structure

3. auricula / auriculae / auriculate
4. synonym of valva. The term is mostly used for spores with extreme valva.

- **auriculate**
  1. ear-shaped.
  2. a leaf or petal with a pair of small, blunt, ear-shaped projections at the base.
  3. ea late (adj) r-shaped, with the leaf base composed of two rounded lobes.

- **autapomorphy**
  derived trait or apomorphy that is unique to only one group or OTU.

- **autocarpous**
  consisting of ripened pericarp alone, having no adnate parts. Applied to fruits that are free from the perianth.

- **autochtonous**
  something that has not been transported.

- **autogamous**
  1. self-fertilised, applied to flowers that are fertilised by their own pollen.
  2. adj., relating to or reproducing by autogamy.
  3. capable of self-fertilisation.

- **autogamy**
  self-fertilisation, the fertilisation of a flower by its own pollen.

- **autotrophic**
  1. nutritionally independent of other organisms.
  2. the ability to use simple chemical elements or compounds to obtain energy for growth.
  3. capable of producing its own food.

- **autotrophic nutrition**
  synthesis of organic food molecules from inorganic compounds such as carbon dioxide.

- **auxiliary cell**
  cell in post-fertilisation development in the Rhodophyta which receives the diploid (2N) zygote nucleus and then develops the gonimoblast filaments

- **auxin**
  1. a group of plant hormones such as indole acetic acid, produced by the tips of growing shoots in plants, that regulate the amount, type and direction of plant growth by stimulating elongation of cells rather than cell division. Auxins diffusing down the stem away from the growing point produce concentration gradients that are responsible for apical dominance and phototropism. At the correct concentrations, auxins induces growth and fruit production even if the

plant hasn't been pollinated (e.g. seedless watermelon). Commercially, auxins are used to promote root growth from cuttings, to promote uniform flowering and to set fruit and prevent premature fruit drop. Synthetic auxins such as 2,4-D and 2,4,5-T stimulate uncontrolled DNA and protein synthesis and are effective herbicides. 2,4-D is still used as a herbicide but use of 2,4,5-T has been banned because it tends to be contaminated with carcinogenic dioxin. Broad-leaved weeds like dandelions are much more susceptible to auxins than narrow-leaved plants like grass and cereal crops. At extremely low concentrations, 2,4-D is used to promote growth of orchid seedlings.

2. an organic substance characterised by its ability to promote growth, particularly of roots. It is an active ingredient in rooting compounds.

3. growth promoting hormones that cause cell elongation and are responsible for many developmental responses including phototropism.

- **auxotonic**

determined by growth. Applied to the movements of plants that are the result of growth, rather than outside stimulation.

- **auxotrophic**

nutritionally independent of organic nutrients.

- **availability**

this term describes the amount of a nutrient in fertiliser or the soil that a plant can immediately absorb; it may be different from the actual amount of the nutrient present.

- **available nutrient**

nutrients found in the soil that can be taken in by plants at rates and amounts that are significant to the plants' growth.

- **awl-shaped**

tapering from the base to a slender and stiff point.

- **awn**

1. a sharp point or bristle, commonly found on grass seed and flower heads. The bristle-shaped terminal appendage, such as the beard of barley, wheat and many grasses

2. (Icel. *ogen*, chaff) a stiff, bristlelike appendage, usually at the end of a structure.

- **axial**

with branches arising from buds in leaf axil.

- **axial bundle**

a major vascular bundle in the shoot or root.

- **axial parenchyma**
  parenchyma cells of the axial system.

- **axial system**
  the vascular elements running vertically up and down a stem, in contrast to the lateral system running horizontally or out to the margin of the stem. The axial system carries water and minerals up the stem in the xylem and carries glucose up and down the stem in the phloem.

- **axial vessels**
  vessels of the axial system.

- **axil**
  1. the upper notch or angle between the axis and any organ such as a Hence, 'auxiliary bud' is a bud in the leaf axil.
  2. the angle formed between a leaf's *petiole* and a stem where an axillary bud develops. The angle formed on an inflorescence, *pedicel* and *rachis* (of an Orchid).
  3. (L. *axilla*, armpit) the angle found between any two organs or structures.
  4. the angle between a leaf or branch and the stem or trunk it springs from.

- **axillary**
  1. pertaining to or growing from the axil of plants; produced in the axil.
  2. (L. *axilla*, armpit) in an axil, growing in an axil, as buds.
  3. in or produced from an axil.

- **axillary bud primordium**
  arises in the axil of a leaf primordium.

- **axillary leaves**
  leaves borne in the axils of branches, as in heterosporous species of *Selaginella*.

- **axis**
  1. see **stem.**
  2. the stem. The central part or upright support of a plant to which organs or parts are attached.
  3. main stem or central line of development on which secondary or side branches are borne.
  4. the central column of an inflorescence or other structure.

- **azonal soil**
  a young soil. A soil having very little or no soil profile development (layers not formed). A soil lacking a well-developed horizon.

- **bacciferous**
  producing berries.

- **backcross**
  in hybridisation, the process of crossing a first generation hybrid with one of its own parents to produce a better variety. A plant produced by backcrossing.

- **backfill**
to fill in a planting hole around a plant's roots with a soil mix.

- **bactericide**
an agent that kills bacteria. Specifically, one that kills bacterial plant diseases.

- **baculum / bacula / baculate**
a cylindrical, free standing exine element more than 1μm in length and less than this in diameter. Example Raistrickia saetosa.

- **bait**
any material used to lure insects, usually to a pesticide. Commonly used to control earwigs and snails.

- **baking soada**
sodium or potassium bicarbonate. Sometimes mixed with horticultural oils and applied to plants (usually roses) to prevent fungal diseases, such as black spot and powdery mildew.

- **bale**
a standard bundle of agricultural merchandise such as straw, cotton or paper usually pressed or bound. The volume and weight of a bale depends on the commodity and locality.

- **balm**
an oily, aromatic resin or other substance exuding from trees. Any aromatic substance secreted from plants.

- **balsam**
an oily, aromatic resinous substance exuding from trees.

- **banding**
placing fertiliser in continuous narrow bands and then covering it with soil. Encircling part of a plant (usually the trunk) or a portion of a garden with some type of material that traps, kills or keeps out pests. Examples include sticky materials, poisonous baits and copper stripping.

- **banding trees**
to fasten a barrier around the trunk of a tree to mark the tree for some attention, such as pruning or to discourage infestation by crawling insects.

- **banner**
the upper petal, usually broad and growing straight up, of some flowers. One of the usually erect petals of the inner upright section of an iris. Also called the *standard*.

- **barb**
a stiff bristle or hair terminating an awn or prickle, usually slanted downward or backward.

- **barbet**
bearded or bearing tufts of long, weak hairs. See **canscence, cili-**

ate, cephalium, felt, glochids, hirsute, lanate, lasiacantha, penicillate, pubescent, pubispina, scabrous, villous, woolly

- **barbed**

bearing sharp, spine-like hooks which are bent backwards.

- **barbellate**

(L. *barbae*, beard) provided, usually laterally, with fine, short points or barbs.

- **bark**

1. the protective exterior covering of the roots, stems and branches of woody plants, exterior to the cambium and including an inner layer of secondary phloem.

2. the dry, dead outer covering of woody branches, stems and roots of plants that is very distinct and separable from the wood itself. It includes all tissue outside the true cambium (growth layer between bark and wood).

3. (ME. *barke* AS. *bare*, bark or rind) the outermost covering of trees and some plants. This is composed of the cuticle or epidermis, the outer bark or cortex and the inner bark or fibre.

- **bark ringing**

this method is used to try to force fruit trees and some other woody plants to flower. A complete ring is cut around the trunk below the lowest branch and another ring is cut right below the first. The bark between the rings is removed and the scar should be covered with grafting wax.

- **barleycorn**

a traditional English unit of length equal to one third of an inch. Use of seeds as average units of weight or length was common in societies based on agriculture. In the Anglo-Saxon era, barley was an important subsistence crop and barleycorns were used as units of length. Three barleycorns 'smooth and round laid end to end' were equal to the Saxonynce (inch). Barleycorns or grains, became the basis of all English weight systems.

- **barren**

sterile or otherwise incapable of reproducing .

- **basal**

1. at or referring to the base of any structure.

2. near point of attachment to substratum.

3. of, at or forming a base.

- **basal leaves**

leaves growing at the base or bottom of a plant.

- **basal taxon / group**
  a group near the root of a clade.
- **base**
  a component of nucleotides, the building blocks of DNA. The four bases found in DNA are represented by A, T, G and C, which have the ability to base pair.
- **base pair**
  formed when complementary nucleotides pair by hydrogen bonding. In DNA, the A nucleotide bonds with T and G bonds with C. Base pairs form the 'rungs' of the DNA ladder and the number of base pairs in a strand can be used to describe the length of DNA.
- **basidiomycete**
  fungi including mushrooms, puffballs, rusts and smuts with hyphae divided by septa and that produce spores on a specialised club shaped cellular structure (basidium).

- **basifixed**
  attached at the base or lower end.
- **basifugal**
  the growth of leaves that develop from the base upward.
- **basilaris**
  arising from the base.
- **basin irrigation**
  the irrigation of level areas that are surrounded by ridges or levees to retain the water.
- **basipetal**
  in direction from apex to base
- **bast**
  a woody fiber obtained from the phloem or inner bark used to make hemp, jute, ramie etc.
- **bast bundles / peripheral strands**
  bundles of thick-walled cells parallel to the midrib, as in *Isoetes*.
- **beaked**
  having a beak or something resembling a beak, beak-shaped, rostrate or ending in a beak-like point.
- **beard**
  a crest, tuft or covering of spreading hairs. The awn or bristle-like appendage upon the chaff of grain and other grasses.
- **bed**
  an area within a garden or lawn in which plants are grown.

• **bedding plant**
an ornamental plant grown for its flowers or foliage that is suited by habit for growing in beds or masses.

• **bendiocarb**
a synthetic insecticide used to control a variety of house and garden pests including earwigs and ants.

• **benomyl**
a systemic fungicide used to control a variety of diseases of ornamentals, usually foliar. It is absorbed into the system of the plant to fight off infection. Common name is Benlate.

• **benthic**
1. attached and submerged.
2. attached to substratum, not planktonic.

• **berry**
1. a pulpy or fleshy fruit with numerous seeds embedded in the pulp.
2. a small, pulpy fruit, which usually contains several seeds (i.e. Huckleberry, mulberry etc.). Any simple fruit in which the inside is fleshy, except the outer skin (i.e. Tomato, banana, grape etc.). The dry kernel of certain kinds of grain.
3. (AS. *berie*, berry) any fleshy simple fruit with one or more seeds and a skin, as a tomato, cranberry, banana, grape etc. A several-sided indehiscent fruit with a fleshy pericarp and without a stony layer surrounding the seeds.

• **betalain**
a nitrogen-containing pigment found only in the family group *Centromeres* which includes the *Cactaceae*.

• **biaxial**
having two axial cell rows.

• **bicollateral bundle**
contains one mass of xylem and two masses of phloem, one toward the interior and one toward the exterior.

• **bicrenate**
crenate with the teeth also crenate.

• **biennial**
1. a plant that typically lives two years from sowing, usually flowering in the second year.
2. a plant that needs two seasons of growth to produce flowers and fruit. It produces leaves the first year and blossoms, bears fruit, goes to seed and dies the next.
3. (L. *biennials*, from *biennis*, *bis*, twice and *annus*, year) a plant requiring two years in which to complete its life cycle, the first year growing only vegetatively, the second flowering, fruiting, then dying.

• **bifacial / vascular cambium**
having two 'faces', i.e. a vascular cambium that produces cells on both sides, in seed plants phloem is produced to the outside and xylem to the inside, compare to unifacial (vascular cambium). See **cambium.**

• **bifarious**
divided into two parts and pointing in two ways or arranged in two opposite rows, as leaves that grow only on opposite sides of a branch.

• **bifid**
1. divided into two equal lobes or parts by a space in the center; forked, like a snake's tongue. Divided in two by a deep cleft.
2. (L. *bifudus*, forked from *bis*, twice and *findere*, to cleave, divide) forked.

• **biflagellate**
having two flagella.

• **bifoliate**
having two leaves.

• **bifoliolate**
having two leaflets as in a compound leaf.

• **bifollicular**
having two seed cases or twin pods, especially in milkweed (Asclepias).

• **biforate**
having two pores or perforations. Also biforous.

• **bifurcate**
split in two. Forked.

• **bilabiate**
1. possessing or having the appearance of two lips; the two-lipped center of a flower. (An example is the flowers of Salvia.).
2. adj., having two lips, as a bilabiate corolla of a flower.

• **bilateral**
1. with equal sides e.g. bilaterally symmetric.

- **bilaterally symmetrical**
said of corolla or calyx (or flower) when divisible into equal halves in one plane only. zygomorphic.

- **bilobate**
having or divided into two lobes, as a bilobate leaf.

- **bilocular**
divided into two cells or compartments.
3. having two small, independent chambers.

- **binate**
being double or in couple, growing in pairs, having only two leaflets to a group.

- **binomial name**
is used to describe each species according to the system devised by Linnaeus. The name consists of the capitalised name of the genus followed by the name of the individual species eg. *Euphorbia obesa.*

- **biocoenosis**
life assemblage. An assemblage of fossils that reflects associations characteristic of the community when it was living

- **biodyanmic**
having to do with the dynamic relation between organisms and their environments.

- **biomass**
(Gr. *bios*, life *massein*, to squeeze) weight of all living material in a unit area at an instantaneous time. May be expressed as g/m2, mt/ha or other similar expressions.

- **biopesticide**
a pesticide comprising a biological control agent,which is usually pathogenic to the target species. An example of a commercially produced biopesticide is the bacterium *Bacillus thuringie-nsis* that kills only insects and not other animals. Its wide host range allows it to be marketed as a sprayable control for a variety of insects, but it may also kill unrelated desirable insects.

- **biota**
the plant and animal life of a region or period.

- **biotechnology**
the application of the study of living things to a myriad of processes, such as agricultural production, hybrid plant development, environmental research and much more.

- **biotype**
all the plants in a specific group, all of which resemble one another in some specific way.

- **biovularia**
a former genus of carnivorous plant from South America currently incorporated in Utricularia .

- **bipinnate**
twice *pinnate*, doubly divided leaf.

- **biplicate**
doubly folded. Twice folded together, transversely, as the first new leaves that sprout from a bean.

- **biramous**
possessing two branches. Dividing into two branches.

- **bird netting**
nylon or plastic mesh used as a drape to keep birds out of fruit trees, berry patches or vegetable gardens. Smaller mesh netting can be placed right over strawberries, raspberries and corn, larger mesh netting may be draped over trees, bushes, grapes or over a supporting structure.

- **bireticulate**
a two layered reticulum consisting of a suprareticulum supported by a microreticulate tectum. Examples Entelea arborescence (Tiliaceae), Phyllanthus oppositifolius (Euphorbiaceae), Salvia azurea (Lamiaceae).

- **birimose**
opening by two slits, as the anthers of most plants.

- **bisaccate**
a pollen grain with two sacci.

- **biseriate**
arranged in two series or rows.

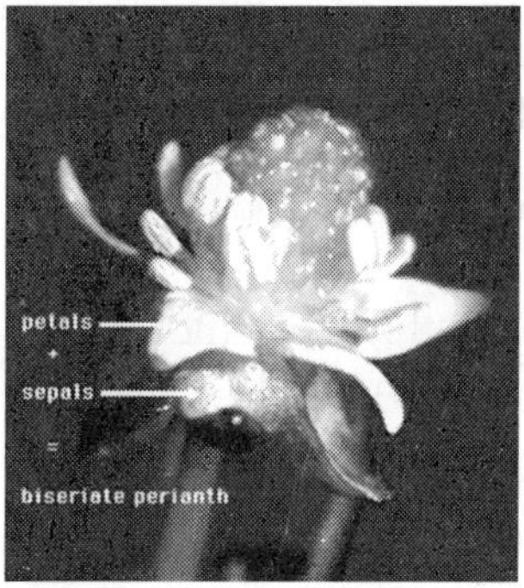

- **bisetose**
having two bristles.

- **bisexual**
1. a flower containing both male and female organs.
2. having both male and female sex organs
3. (L. *bis*, twice *sexus*, sex) having both female and male reproductive organs present and functional in the same flower; hermaphrodite; amphisporangiate. Said of a plant having all bisexual flowers.

- **bispinous**
having two spines.

- **bisporangia**
  sporangia that divides to produce 2 spores perhaps by mitotic division, in Corallinales and some other Rhodophyta.

- **black plastic**
  sheet plastic that is placed around plants as a mulch to obstruct weeds, hold moisture and absorb heat. Also called black polyethylene or black poly sheeting.

- **bladder**
  a hollow chamber designed for trapping or flotation, notably in bladderworts.

- **bladderwort**
  any plant of the genus Utricularia .

- **blade**
  1. the expanded portion of a leaf or petal.
  2. a plant's leaf, such as a blade of grass. The expanded part of a leaf, petal, sepal etc. as distinguished from the petiole or leaf-stalk.
  3. the expanded portion of a leaf.
  4. flat, leaf-like structure.
  5. (AS. *bladed*, leaf) the leaf of a plant, especially grass. The flat or expanded portion of a leaf, lamina.

- **blanch**
  to whiten or prevent from becoming green by excluding light. Blanching is applied to the stems or leaves of plants such as celery, lettuce and endive. This is done by banking up the soil around the stems of the plants, tying the leaves together to keep the inner ones from light or covering with pots, boxes etc.

- **blastema**
  the point of growth from which any organs or part of an organ is developed.

- **bleeding**
  when sap or *latex* oozes from a plant or tree.

- **bloom**
  (ME. *blome*, a blossom) a blossom. The flower of a plant. an expanded bud. The opening of flowers in general, leaves, flowers or fruits.

- **blossom**

(ME. *blossome*, a flower) a flower or bloom, esp. of a fruit bearing plant. A state or time of flowering, literally and figuratively.

- **blowball**

the downy heads of seeds produced after the flowers have died. (i.e. dandelion, salsify etc.).

- **bog**

(Ir. *bogach*, a bog, from Gale. *bog*, soft moist) a quagmire covered with grass or other plants, wet, spongy ground. A small marsh, plant community on wet, very acid peat.

- **bog garden**

a garden devoted to moisture-loving plants. Low-lying, marshy sites are good for this.

- **bole**

the trunk of a tree.

- **bombycinus**

silky.

- **bone meal**

ground bone, raw or steamed, that is used as a fertiliser to add phosphoric acid to the soil. It is especially beneficial in herb gardens.

- **bonsai**

a shrub or tree grown in a container and dwarfed by pruning, pinching and wiring to produce a desired shape.

- **bonsai dish**

a planter for holding a miniaturised tree or shrub.

- **border**

a narrow strip of garden that encloses or separates a portion of lawn or other landform by a path, walk, driveway, street or building. Usually planted with perennials or flowering shrubs for spring and summer bloom.

- **border fence**
  a short fence, usually only 12 to 18 inches high, of wood or metal, used to mark and separate flower beds from foot traffic and lawn mowers.

- **boreal**
  northern, colder waters. A climatic zone bordering the Arctic or Antarctic.

- **boric acid**
  a salt that can be effectively used as a barrier against some crawling household pests such as ants. If kept dry, it can be used outdoors.

- **boron**
  a trace element usually present in most soils but sometimes added by applying fertiliser.

- **bosket**
  a grove. A thicket or small plantation in a garden, park etc., formed of trees, shrubs or tall plants.

- **botanic**
  pertaining to botany or the scientific study of plants.

- **botanical insecticide**
  any natural insecticide derived from plants. They include pyrethrum, neem, rotenone, rynia, sabadilla and nicotine sulfate. They break down quickly when exposed to sun and must be applied often. They are usually non-toxic, but not always.

- **botany**
  the science of plants, their classification and study.

- **botryoid**
  having the form of a bunch of grapes.

- **bottomland**
  lowlands along streams and rivers, usually on alluvial floodplains that are periodically flooded.

- **botuliform**
  sausage shaped, long and cylindrical with each end curved inwards.

- **bough**
  an arm or branch of a tree.

- **bow saw**
  a tool having a straight blade attached at both ends to a curved handle, in all making the shape of the letter D. It is used for cutting branches.

- **brachy-**
  a prefix for short. Synonym of brevi.

- **brachysclereids**

  also called stone cells, length and width nearly equal.

- **brackish**

  a mixture of salt and fresh water with intermediate salinity.

- **bract**

  1. a reduced or modified leaf subtending an axis. Often found at the base of a flower or flower cluster and may form a cup around the flower in e.g. *Compositae* or may be brightly coloured and resemble petals in e.g. *Poinsetta.*

  2. a modified leaf at the base of a flower or flower cluster. It may resemble a normal leaf, be reduced and scale-like in appearance or large and brilliantly coloured.

  3. (L. *bractea*, a thin metal plate) a modified leaf, growing at the base or on the stalk of a flower. It usually differs from other leaves in shape or colour.

  4. a modified leaf providing an axil for a flower or inflorescence to grow from or found on a flowering stem.

- **bracteate**

  having bracts.

- **bracteolate**

  furnished with bracteoles.

- **bracteole**

  (NL. *bracteola*, from L.) a thin gold leaf a small bract; especially one on a floral axis. pl. bracteoles.

- **branch**

  a stem growing from the trunk of a tree, shrub or other plant (the smaller offshoots are called branchlets, twigs or shoots), a bough.

- **branchlet**

  1. a small usually terminal branch.

  2. smallest pigmented branches or secondary branches, ramuli.

- **breed**

  a group of plants, developed by man, which will not keep their characteristics in the wild.

- **breeding**

  the directed development of better crops through the selection of individuals expressing the desired traits. Breeding alters the genetic makeup of the population and this may be considered a form of genetic modification of that population. The alteration of plant characteristics through breeding was accomplished much earlier than the discovery of modern genetics.

- **brevi-**
  a prefix for short. Example brevicolpate.

- **breviaxe / breviaxal**
  pollen grains with a polar axis that is shorter than their equatorial diameter. Synonym of oblate (s.l.).

- **breviaxy**
  a phyletic series of pollen forms ranging from subspheroidal tricolpate, to tricolporate, to oblate tricolporate. See **successiform**.

- **brevissimi-**
  a prefix for very short.

- **bridge**
  a feature in which the margins of the colpi are raised in the equatorial region and connected with each other, forming a bridge over the ectocolpus and dividing it into two parts. Example Hemandradenia (Connaraceae).

- **bristle**
  a stiff, sharp hair on a plant.

- **broad-base terrace**
  a low embankment that is constructed across a slope to reduce runoff and/or erosion.

- **broadcast**
  scattered upon the ground with the hand, as in sowing seed, instead of sowing in drills or rows.

- **brocchinia**
  a small genus of bromeliads of which two species to date have been found to be carnivorous.

- **brochus / brochi / brochate**
  a brochus consists of one lumen of a reticulum and half of the width of the adjacent muri.

- **brown forest soil**
  a soil rich in humus with a dark brown surface soil.

- **brown podzolic soil**
  a mixed humus and mineral soil with a thin mat of decayed leaves; usually found under a deciduous (leafy) or mixed deciduous and coniferous forest.

- **brown soil**
  a brown surface soil found under grasses and shrubs that gets lighter in colour the deeper you dig.

- **brush**
  the small trees and shrubs of a wood, a thicket or small trees, scrub.

- **bud**

1. an unopened flower or a growing tip surrounded by its immature perianth segments or leaves.

2. A growing point enclosed by closely overlaid rudimentary leaves. Buds may contain foliage or flowers. Bulbs and bulbils are forms of leaf buds. Flower buds are unexpanded blossoms.
3. to graft a bud of 1 plant on the stem of another (i.e. to bud a garden rose on a brier).
4. (ME. *budde*AS. *budda*, beetle) a small swelling or projection on a plant, from which a shoot, cluster of leaves or flowers develops. A rudimentary, undeveloped shoot, leaf or flower, gemma.
5. a small cell-mass in the process of dividing from a stem to form a flower, leaf or another stem.

- **bud imprint**

the outline of the margins, teeth or other features of one leaf impressed on another leaf while the two leaves are pressed together in the bud and remaining as a permanent marking after the leaves become separated. This effect is particularly marked in certain species of *Agave*.

- **bud mutation**

a bud variation resulting from local genetic alteration and producing a permanent modification that usually can be retained by grafting.

- **bud sport**

the product of bud mutation or bud variation.

- **bud variation**

in the outgrowth of a bud, the deviation in any respect from the ordinary growth of the plant, producing what is commonly known as a sport. Many varieties arise this way, such as Red-Delicious Apples and these are perpetuated by bud propagation.

- **budding**

1. the production of buds.
2. a form of propagation in which a single bud is grafted to the stock. Also called bud grafting.

- **budding strip**
  a rubber strip, typically 1/8th to 3/8th inches wide and up to 8 inches long, used to hold grafts.

- **buffer**
  a substance in the soil that will chemically act to resist changes in the soil's reaction or pH, usually clay or fine organic matter.

- **buffer strips**
  strips of perennial grass or other erosion-resisting vegetation.

- **bulb**
  a true bulb is a complete package containing next year's flower already forming inside. If you were to cut a tulip, daffodil, hyacinth or amaryllis bulb in half, you would see a complete embryonic plant inside, with tiny flowers, stems, leaves and roots. The contents of the bulb are often enclosed in protective, fleshy scales, which are held together by a small basal plate. The scales are modified leaves that contain enough nutrients to sustain the plant through dormancy and early growth. They may be loose and open like those of a lily or tightly closed like those of a hyacinth. In many bulbs, a paper-thin tunic protects the scales (lilies don't have a tunic). Roots will grow from the bulb's basal plate. The foliage of bulbous plants shouldn't be cut off while green. It should be allowed to die naturally because it supplies the food for the next season's growth. There are several different ways in which bulbs reproduce. Some reproduce by a process called annual replacement. The old bulb, called the 'mother' bulb, dies after flowering and is replaced by new 'daughter' bulbs of various sizes. When the time is right, the largest of these will flower the following season and the others will bloom in the years to come. Other bulbs reproduce by offsets. The mother bulb lives for two to three years and the new bulbs (offsets) are produced alongside the mother bulb. Offsets can be removed and planted elsewhere in the garden.

- **bulb planter**
  a sharp-edged, tapered cylinder used to remove a plug of soil or sod, in which a bulb is placed. The plug is then returned to the hole to cover the bulb.

- **bulbil**
  a small vegetatively derived plant produced on an inflorescence.

- **bulblet**
  a little or secondary bulb, specifically, a small aerial bulb or bud with fleshy scales, growing in the axils of leaves, as in the tiger lily or taking the place of flower buds, as in the common onion. Also called bulbil.

- **bulbous**
  swollen.

- **bullate**
  1. blistered or puckered.
  2. having a pimpled or blistered surface.

- **bundle sheath**
  the layer of tightly-packed cells that surround the vascular tissue in leaves.

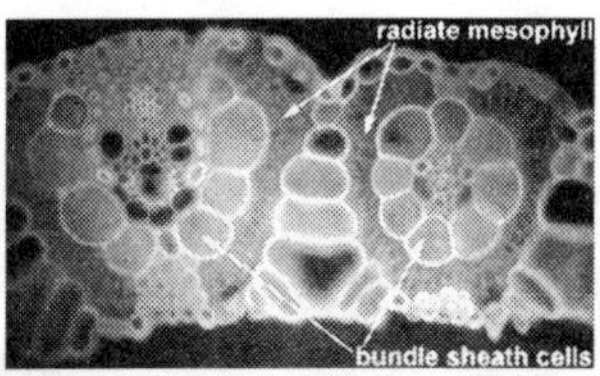

- **bundle sheath extension**
  a group of cells that connect a vein to the epidermis in a leaf.

- **bur**
  the rough, prickly covering of the seeds of certain plants, such as the chestnut or burdock. Also spelled burr.

- **burl**
  a hard, woody growth on the trunks or branches of trees that is sometimes used for making bowls, veneer and coffee tables, such as from the redwood.

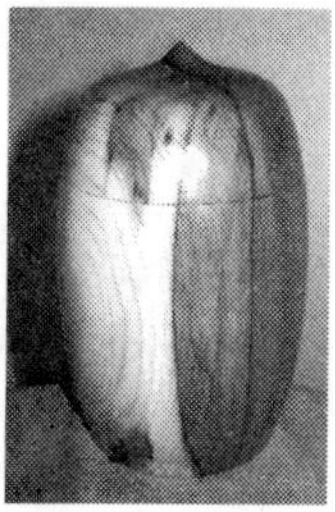

- **burlap**
  a loosely woven fabric made of jute or hemp used to protect newly seeded lawns from wind, water and birds.

- **burn lime (CaO)**
  a caustic solid used to neutralise acid soil or to raise the pH of the soil.

- **bursiculate**
  resembling a small pouch or having a small, pouch-like cavity.

- **bush**
  1. a thicket, a clump of trees or shrubs.

2. a low and multi-branched shrub.
3. an expanse of forest or shrubby vegetation.
4. a low, often woody, plant normally branching from the base, rather than having a single stem that branches higher up. Also applied to a thicket or to an area covered with trees, bushes or scrubby vegetation in e.g. Australia or South Africa or generally to any wilderness area.

- **bushel**

an English Imperial measure of volume, of a container or basket used to measure such a volume, typically used to measure dry goods such as grain or fruit. Such a cylindrical container would be 18.5 inches in diameter and 8 inches deep. A bushel = 4 pecks, 8 English gallons (defined as the Winchester bushel by King Edward 1 in 1303), 1.2844 cubic ft., 2219.36 cubic inches (Metric: 36.369 liters). A heaped bushel is traditionally 27.8 percent larger than a regular or struck bushel which has been leveled. Agricultural goods such as grain are traditionally sold by the bushel, but as they tend to settle during shipping disputes over volumes delivered can arise. Therefore bushels of a particular commodity are defined as standard weights, often enshrined in law and it is really the weight of goods that is sold rather than the volume.

- **butterfly garden**

a garden planted for the purpose of attracting butterflies by using plants whose flowers produce nectar for them. (i.e. butterfly bush -Buddleia and butterfly flower - Asclepias).

- **buttonbush**

a common name for *Cephalanthus Occidental*, a North American shrub, on account of its globular flower heads.

- **cactus**

this is the common name for plants in the group, Cactaceae. All cacti are *succulent*, although not all succulents are cacti. Cacti can be distinguished from other succulents by their *areoles* or pad-like buds, from which shoots, flowers and spines grow. Areoles aren't always easily seen.

- **caducous**
falling off early. (L. *caducus*, falling, from *cadere*, to fall) said of a plant part, such as a sepal, petal or leaf, that falls off quickly or early.

- **caespitose**
1. forming a cluster or clump of stems by repeated branching of the stem above ground, at or near the base of the plant. Of plants that offset freely to form a large clump.
2. tufted; growing in small clumps.

- **caffeine**
a methyl-xanthine alkaloid occurring naturally in some plants used to make beverages and foods including *Coffea arabica* berries (coffee), *Thea sinensis* leaves (Tea), *Cola acuminata* nuts (Coca Cola), *Themobroma cacao* beans (chocolate) and added to a wide range of manufactured foods. Caffeine is one of the most widely used mood-altering drugs and has probably been used through the whole of human history.

- **calcareous**
chalky, derived from limestone or with an excess of lime. Calcified, hardened with lime (calcium carbonate), e.g. in the Corallinales

- **calcareous nannofossils**
nannofossils largely composed of calcium carbonate. Many forms belong to the Coccolithophyceae.

- **calcareous soil**
soil containing carbonate of lime or limestone, chalky or limy.

- **calcifuge**
a plant unable to stand chalky (limy) soil.

- **calcium**
an element needed by plants. It can be added to the soil.

- **calcium carbonate**
a salt CaCo3 found in plant ashes, bones, shells, calcite and aragonite.

- **calcium chloride**
a deliquescent salt CaC12 used as a drying and dehumidifying agent.

- **calcium cyanamide**
a compound CaCN2 used as a fertiliser and weed killer.

- **calcium phosphate**
1. a phosphate used as a fertiliser.
2. naturally occurring phosphate of calcium occurring as the main constituent of phosphate rock.

- **caliche**
a term used to identify cemented deposits of calcium carbonate.

• **callus**

1. a tissue composed of large thin-walled cells, usually produced in response to injury.
2. any unusually hard, thickened area or swelling on a plant, such as the thickening that is formed over wounds (such as the base of a cutting) by which the inner tissues are protected and healing is effected.
3. a waxy or fleshy protuberance, as on a *labellum* (of an Orchid).
4. an irregular proliferation of cells.
5. abnormal, thickened part often a reaction of wounding
6. (L. *callus, collum*, hard skin) a hard protuberance or callosity, new tissue covering a wound.

• **calymmate**

describing tetrads or polyads in which the sexine/ectexine of each monad is well differentiated and forms a continuous envelope around the unit. Example Acacia (Mimosaceae).

• **calyptra**

a hood or covering on a flower or fruit. Also called calypter.

• **calyptrate**

furnished with a calyptra, as a capsule or a flower; resembling a calyptra, as a calyx that comes off like a lid.

• **calyptrogen**

the cell layer from which the root cap originates.

• **calyx**

1. a small whorl of modified leaves (sepals) at the base of a flower and, where present, enclosing the other parts of the flower in bud. Th ese may occur individually or joined into a cup or tube and may be of any colour, but commonly green.
2. the outer part of a flower, usually small and green but sometimes showy and brightly coloured; it is formed from the sepals and encloses the petals in a bud.
3. (Gr. *kalyx*, a calyx, cup) the outer covering of a flower external to the corolla, which it encloses and consisting of a whorl of leaves or sepals, usually of a green colour and less delicate in texture than the corolla.
4. a protective structure around a flower formed by the sepals collectively.

- **calyx tube**
  tube formed by wholly or partially fused sepals. Not the floral tube of an epigynous or perigynous flower.

- **cam metabolism - crassulacean acid metabolism**
  a metabolic adaptation allowing temporary storage of carbon-dioxide as C-4 organic acids (e.g. malate, aspartate) and later release of carbon-dioxide, by decarboxylation of these acids, for fixation into sugars. This allows uptake of carbon-dioxide during the cool nights, when water loss by transpiration is relatively low, while providing a store of carbon-dioxide for photosynthesis during the day. C-4 metabolism is typical of *Agaves, Cactaceae, Crassulaceae* and tropical grasses and can supp ort high rates of growth under optimum conditions.

- **cambium**
  1. a layer of tissue giving rise to secondary growth in stems and roots by cell division. Seasonal growth within the cambium produces annual growth rings. The corky bark of the cork oak tree is formed from cambium cells.
  2. a lateral meristem that produces secondary growth.
  3. a layer of tissue, one cell thick, formed between the wood and the bark of vascular plants that is capable of developing new cells. On one side, cambium develops into a layer of new wood and on the other side, into bark. At the same time, new cambium is produced. This process happens every year and increases the size of the trunk, as indicated by its *annual rings* when wood is cut into a cross section.
  4. (L. *cambiare*, to exchange, more at change) the layer of tissue between the bark and wood in woody plants, from which new wood and bark develops.

- **camera / camerae / camerate**
  a cavity formed by the separation of tw o wall layers in spores that lacks an infrastructure.

- **campaniform**
  bell-shaped, resembling flowers of the genus *Campanula*.

- **campanulate**
  1. bell shaped, applied to the shape of a flower.
  2. bell-shaped, resembling flowers of the genus *Campanula*.

- **canaliculate**
grooved or channelled longitudinally.

- **canals**
as in *Equisetum*.

- **cancellate**
(L. *cancellatus*, pp., of *cancellare*, to make like a lattice) latticed or resembling a latticed construction, usually said of a surface such as that of an achene or seed.

- **cane**
a long and slender, jointed, rigid, woody stem that is hollow or pithy, such as some grasses and palms and rattan, bamboo and sugarcane; the stem of raspberries or blackberries.

- **canescence**
a dense covering of very fine white or gray short hair. See also: **barbate, ciliate, cephalium, felt, glochids, hirsute, lanate, lasiacantha, penicillate, pubescent, pubispina, scabrous, villous, woolly.**

- **canescent**
growing white or gray. Applied to plants with grayish-white, short, downy leaf hairs.

- **canker**
a lesion or an oozing point of decay in the tissue of a plant, usually on the trunk or branches. Caused by wounds, disease or insects.

- **canola oil**
a low-erucic acid rapeseed oil valued at $2 billion per year to the Canadian economy. The first modified vegetable oil to achieve large-scale commercial production, canola was developed because of nutritional concerns in animals with the fatty acid, erucic acid. In canola oil, oleic acid replaced the erucic acid found in traditional rapeseed oil.

- **canopy**
the foliage cover of a plant community especially woodland.

- **capillary**
1. water movement within the soil.
2. water held by or found in the soil resulting from surface tension.
3. (L. *capillaris*, from *capillus*, hair, from *caput*, head) resembling hair in the manner of growth; very slender, threadlike.

- **capillary attraction**
water moving against gravity and moving up to the soil's surface.

- **capillary matting**
an interwoven fabric that wicks water from a reservoir. Used in propagators, seed flats, pots or

English baskets to provide plants with a constant supply of moisture.

- **capillary porosity**
  the volume of small pores in the soil that holds water.

- **capilliform**
  hair-shaped, finer than filiform.

- **capillus / capilli / capillate**
  synonym of fimbria.

- **capitate**
  1. head-like, collected in a dense cluster.
  2. (L. *caput*, head) enlarged or swollen at tip, gathered into a mass at apex, as compound stigma. A knoblike stigma terminating a style.
  3. terminated by an enlarged and rounded head.

- **capitulum**
  1. a dense head of flowers.
  2. (L. *capitulum*, small head) an inflorescence forming a head of sessile flowers or florets crowned together on a receptacle and usually surrounded by an involucre.

- **cappa / cappae**
  1. the thick-walled proximal side of the corpus of a saccate pollen grain.
  2. the thin-walled distal side of the corpus of a saccate pollen grain.

- **capsula / capsulae / capsulate**
  1. an outer structure of a spore projecting at the equator and completely enclosing the spore body.
  2. a multi-chambered seed pod with many seeds per chamber that opens when dry and ripe.
  3. a pod or seed vessel made of two or more cells, which becomes dry and splits open when mature to release its seeds.
  4. (L. *capsula*, a little chest) a case, pod or fruit, containing seeds, spores or carpels. It usually bursts when ripe.

- **captan**
  a man-made fungicide used to control many disease such as leaf spots and damping-off.

- **caput / capita / capitate**
  the expanded apex (head) of a columella.

- **carbaryl**
  a man-made insecticide that kills insects (even beneficial ones) on contact. It is very poisonous to bees. Commonly called Savin.

- **carbon**
  one of the most common chemical elements found in the soil and air that is needed by living plants.

- **carbon dioxide (CO_2)**
  a colourless gas found in the air.

- **carbonate**
  a salt or ester of carbonic acid (H_2CO_3). Carbonates can combine with other elements to form minerals, for example with calcium (calcium carbonate) or with iron (iron carbonate).

- **carbon-nitrogen ratio**
  the ratio of weight of organic carbon to the weight of nitrogen found in the soil.

- **carinal canal**
  1. a canal in the xylem of some sphenophytes, for example Equisetum, that results from extension and rupture of the protoxylem elements.
  2. a canal beneath a stem ridge associated with a vascular bundle.

- **carinate**
  (L. *carinatus*, from *carina*, a keel) shaped like the keel of a ship having a longitudinal prominence on the back, like a keel, applied to a calyx, corolla or leaf.

- **carotenoid - carotene**
  a class of violet to orange or yellow fat-soluble unsaturated hydrocarbon pigments, usually with 9 conjugated double bonds, found in most higher plants and responsible for the colour of carrots, squashes and sweet potatoes. Carotene assists photosynthesis by transmitting energy from absorbed light to chlorophyll. Dietary carotene is converted to Vitamin A in the liver of animals, which are unable to synthesis their own and becomes concentrated in fat and cell membranes where it acts as an antioxidant. Carotene may cause a yellowish colour in people who consume large amounts in their diet. Carotene pigments extracted from plants are used to colour foods such as cheese and margarine.The oxygenated derivatives of carotenes are known as xanthophylls.

- **carpel**
  1. female reproductive part of flower including ovary, style and stigma, which may be solitary, grouped or fused.
  2. a simple *pistil* or one of several members composing a compound pistil or fruit. In other words, it is the organ of a plant that bears *ovules*.
  3. a unit of the pistil consisting of a single modified leaf, a simple pistil consists of one car-

pel and a compound pistil consists of fused carpels.
4. (Gr. *karpos*, fruit) a simple pistil, regarded as a modified leaf also, any of the two or more carpels that unite to form a compound pistil; the unit of structure of the female portion of a flower.
5. collectively the stigma, style and ovary; the female reproductive parts of a flower.

- **carpogenic**
fruit-producing.
Also carpogenous.

- **carpogonium**
haploid (N) female sexual structure in Rhodophyta consisting of single cell and its extension, the trichogyne.

- **carpophore**
1. the wiry stalk that bears the carpels of some compound fruits.
2. (Gr. *karpophorus*, bearing fruit; *karpos*, fruit and *pherein*, to bear) generally the organ that supports the carpels; specifically, a very much elongated axis to which the carpels are attached.

- **carposporangia**
single celled structures that produce t he diploid (2N) carpospores on the parasitic carposporophyte in Rhodophyta.

- **carpospore**
non-motile diploid (2N) spores formed on parasitic carpospor-ophyte stage of the Rhodo-phyta, which germinate to form tetrasporophyte stage.

- **carposporophyte**
the diploid (2n) stage of Rhodophyta which develops after fertilisation of the carpogonium.

- **cartilaginous**
fleshy, but firm or even tough and leathery.

- **caruncle**
an outgrowth on a plant or animal such as a fowl's wattle or a protuberance near the hilum of certain seeds.

- **caryopsis**
a small, one-seeded, dry fruit in which the fruit and seed are incorporated into a single grain, as in wheat and all other cereal grains.

• **casparian strip**
a band of suberin within the anticlinal walls of endodermal and exodermal cells.

• **castanea**
(L., a chestnut, from Gr. *kastanon*) a genus of trees typified by the common chestnut.

• **castaneous**
relating to or having the colour of a chestnut.

• **cata**
a prefix indicating the location of features, such as apertures, on the proximal face.

• **catadromous**
having the first lobe or segment of a pinna arising acroscopically in compound leaves.

• **catalase**
an enzyme capable of decomposing hydrogen peroxide into water and oxygen.

• **catalyst**
something that will increase the rate of a chemical reaction.

• **cataphyll**
1. these are the rudimentary leaves that precede a stage of growth. (such as the *cotyledons* of an embryo, the scale-like leaf of a rhizome, bud or at the base of a stem).
2. (L. from *cata* and *-phyll*) any rudimentary leaf, as a bud scale, preceding the true foliage leaves.
3. a leaf modified to protect a dormant bud.

• **cataphyllary leaves**
rudimentary or scale-like leaves which act as a covering of buds.

• **catazonasulculate**
describing a pollen grain with a ring-like sulculus situated between the equator and the proximal pole.

• **catch crop**
a method of increasing garden productivity by filling in the empty spaces created when slower-growing vegetables are harvested with fast-growing crops. (i.e. green onions or radishes are grown in the space created when a broccoli plant is cut.)

• **catena**
the soil that is formed from similar parent materials but doesn't have like soil characteristics because of the differences in drainage or relief.

• **cation exchange**
the exchange of cations, held by a soil absorbing complex, with other cations.

• **cation exchange capacity**
the measurement of the total amounts of exchangeable cat-

ions that can be held by the soil. Expressed as milliequivalent per 100 grams of soil.

- **catkin**
  1. a dense, scaly spike of flowers that fall after flowering or fruiting, as in the willow or birch; an *ament*.
  2. (L. a dim. of *cat*, from its resemblance to a cat's tail) a scaly spike, the flowers of which are unisexual and petalless.

- **catopsis (g)**
  a genus of bromeliads of which only one species has been found to be carnivorous.

- **caudate**
  having a tail-like appendage, a 'tail' or narrowed, apical extension of some sepals and petals.

- **caudex**
  1. a swollen base to the stem used as a storage organ.
  2. the swollen water-storage tissue, usually consisting of both the root and stem of a succulent or woody plant.
  3. (L. *caudex*, stem of a tree) the base of a perennial plant. The axis or stem of a woody plant, especially of a palm or tree fern.

- **caudiciform**
  possessing a *caudex*. A plant with a swollen stem.

- **caudicle**
  extensions of tissue derived from the anther and connected to *pollinia* in many Orchids.

- **caulescent**
  1. having an obvious stem that grows above the soil's surface.
  2. (L. *caulis*, a stem and -*escent*) having a well-developed stem above ground level. (usually of flowers) more or less stemmed; appearing to have a stem.

- **cauline**
  (L. *caulis*, stalk or stem).

- **caulocarpous**
  bearing a fruit repeatedly on the same stem, applied to plants that have perennial stems.

- **cavate**
  in spore terminology, a synonym for camerate.

- **cavea / caveae / caveate**
  a cavity between two layers of the exine extending to the colpus margin where the layers meet. Example Ambrosia (Compositae).

- **caverna / cavernae**
  synonym of fastigium.

- **cavium**
  a subpolar chamber formed at the end of three anastomosing cavernae.

- **cavum / cava / cavate**
  synonym of cavea.

- **CDNA (complementary DNA)**
  DNA produced in vitro in such a way that the sequence of bases is complementary to a particular messenger RNA. Complementary DNA (cDNA) is used to study gene expression because it is more stable than RNA and is more amenable to recombinant cloning techniques.

- **cell**
  1. the basic structual unit, which may be capable of replication or differentiation, from which all living tissues are made.
  2. a small, usually microscopic mass of protoplasm, usually with one or more nuclei and some nonliving material, with an external membrane. The basic structural and functional unit of life.
  3. the structural unit of an organism, a cavity.

- **cell wall**
  a rigid layer of cellulose and other polysaccharides, proteins and sometime lignin on the outside of the plasma membrane of a plant cell.

- **cells**
  the smallest living unit of an organism. The outer boundary of a plant cells is defined by a rigid cell wall.

- **cellulose**
  1. the major constituent of plant cell walls and probably the most abundant organic substance on earth with about 10^{11} tons synthesised each year. Cellulose consists of beta-D-glucose units linked together with beta-1,4' linkages (compare with starch) into very long, high molecular weight polysaccharide molecules (typically up to several million kDa). Up to 50% of the carbon content of plants may be cellulose.
  2. polysaccharide that consists of a long unbrached chain of glucose units; cellulose is the main constituent of the cell walls of most land plants.
  3. the structural (microfibrillar) portion of the plant cell wall. Cellulose is a polymer of glucose.
  4. (Fr. from L. *cellula*, dim. of *cella*, a small room) the chief substance composing the cell walls or woody part of plants. A carbohydrate of unknown mo-

lecular structure but having the composition represented by the empirical formula $(C_6H_{10}O_5)_x$.

- **central**
positioned at or near the centre of an area, as opposed to being peripheral in position. A spine originating in the centre of an areole as opposed to those growing around the edge of the areole.

- **central canal**
the large centrally located air space in the stem.

- **centrifugal**
developing from the centre outward.

- **centripetal**
developing inward toward the centre.

- **centrosymmetrical**
synonym of radially symmetric.

- **centrum**
(L. from Gr. *kentron*, centre) central portion.

- **cephalium**
1. modified flowering zone of stems of certain species of cactus, often ornamented by copious development of hairs or bristles. In some species (e.g. Melocactus) stem growth ceases once the cephalium is produced. In others, a pseudocephalium is produced while stem growth continues. see **barbate, canescence, ciliate, felt, glochids, hirsute, lanate, lasiacantha, penicillate, pubescent, pubispina, scabrous, villous, woolly.**
2. a densely spined crown, which produces flowers and fruit. It is produced only by species Discocactus and Meloc-actus.

- **cephalotus (g)**
a monotypic genus of pitcher plant from SW Australia, the Albany Pitcher Plant.

- **ceraceous**
waxy. Applied to bodies that have the texture and colour of new wax, as the pollen masses of many orchids.

- **cereus**
wax-like, candle-like.

- **cernuous**
drooping, pendulous, having the top curved or bent down.

- **cespitose**
growing in tufts or clumps, matted.

- **chaff**
1. the glumes, husks or other pieces of the leaves and stems of wheat, oats or other grains and grasses, especially when separated from the seed by threshing and winnowing.

2. (AS. *ceaf*, chaff) dry scales or bracts, as those on the receptacle subtending the flowers in the heads of certain Compositae.

- **chalaza**
  the part of the ovule where the seed stalk and the outer coverings attach to each other.

- **Channelled**
  1. hollowed out; trough-like, applied to petioles, leaves etc.
  2. having a deep longitudinal groove.

- **charcoal**
  an amorphous form of carbon produced by partially burning or oxidising wood or other vegetable material in fires from which air is partially excluded by covering with cut turf or in large kilns in the absence of air. Uses include drawing material, as a fuel, filter, gas absorbent etc.

- **chartaceous**
  1. of or like paper or parchment. Usually not green.
  2. (L. *chartaceus*, from *charta*, a leaf of paper) having the texture of thin but stiff paper.

- **chasmogamous**
  requiring opening of the flower for pollination.

- **chasmogamy**
  (Gr. *chasma*, an opening, chasm and *gamos*, marriage) the opening of the perianth of a flower for the purpose of fertilisation; contrast with cleistogamous.

- **chemical drenches**
  can be used to control soil-borne disease.

- **chemical gardening**
  a method of cultivating plants in which chemicals are used to control undesirable pests or weeds.

- **chemonastic**
  regarding the chemically-induced movement of plant parts.

- **chemotropism**
  bending, growth or movement of a part under the influence of a concentration gradient of chemical substances. Aerotropism- special case of a response to oxygen.

- **chimera**
  1. the combination of tissues of different genetic origin in the same part of a plant. Examples of chimeras include some variegated cultivars of *Sansevierias*, although there can be many reasons for variegation.
  2. a plant consisting of two or more genetically distinct tissues or an artificially produced plant having tissues of several species.

- **chipper**

a tool made of a chute, blades and an electric or gas motor; used for turning branches into wood chips.

- **chitin**

1. may be the second most abundant polysaccharide in nature, after cellulose to which it is structurally related, but see also lignin. At least 10^9 tons of chitin are thought to be synthesised each year. Chitin is found in the cell walls of many lower plants such as yeast, mushrooms and other fungi. It also forms a substantial part of the exoskeleton shells of crustaceans, such as crabs shrimps and lobsters, as well as in the exoskeletons of marine zooplankton, including copepods and in corals and jellyfish. Insects, such as butterflies and ladybird beetles, have chitin in their wings. Chitin mainly consists of the aminosugar N-acetyl-D-glucosamine linked together with beta-1,4' linkages, which may be partially deacetylated to a form of chitin called chitosan. Chitin is usually found complexed with other polysaccharides and proteins.

2. a natural pesticide made from pulverised crustacean shells. This is added to the soil to increase the feeding of soil organisms on nematode eggs, which helps control these pests.

3. the polysaccharide coating of microforaminifera, scolecodonts and the spores and hyphae of certain fungi. Highly resistant to forms of deterioration other than oxidation.

- **chitinozoan**

1. marine microfossil of uncertain affinity generally shaped like a flask, occurring individually or in chains. Usually assumed to be animal remains.

2. marine microfossils of uncertain affinity generally shaped like flasks, occurring individually or in chains. stratigraphic range from uppermost Cambrian to Devonian. Usually assumed to be animal remains.

- **chloranthus**

with green flowers.

- **chlorenchyma**

contains chloroplasts and functions in photosynthesis.

- **chlorophyll**

the green mettle-pigment in plants concerned with harvesting energy from light and transmitting the energy to an ordered chain of enzymes, ultimately responsible for the production of sugars. Each Chlorophyll molecule contains a central magnesium atom chelated by four nitrogens from a large porphyrin ring and the two forms may be distinguished by substitution of a methyl group in Chlorophyll a for an aldehyde in Chlorophyll b.

- **chloroplast**

1. a specialised cytoplasmic organelle containing chlorophyll, characteristic of plants and responsible for using energy captured from visible light to fix carbon from carbon-dioxide and synthesise carbohydrates as energy storage molecules. Like mitochondria, chloroplasts contain a small amount of DNA which encodes genes for a few chloroplast proteins.
2. contains chlorophyll, internal membranes organised as grana, specialised for photosynthesis.
3. double-membrane bound organelle inside the cell with photosynthetic pigments on thylakoids.

- **chlorosis**

(Gr. *chloros*, pallid) abnormal condition characterised by absence of green pigments in plants.

- **chlorothalonil**

a multi-purpose fungicide used to treat a variety of plant diseases. Commonly known as Daconil.

- **cholla**

any member of the genus *Opuntia* with short cylindrical-stems.

- **choripetalous**

having the petals unconnected.

- **chorisis**

the multiplication, by congenital division during development, of a leaf or floral organ that is normally whole.

- **chromoplast**

1. a specialised cytoplasmic organelle containing pigments other than chlorophyll including yellow or orange carotenoid pigments.
2. contains red orange or yellow carotenoid pigments, impart colour to fruits etc.

- **chromosome**

1. tiny condensed X or Y-shaped sub-cellular bodies seen at the time of cell division, containing the genetic material (DNA) in the cell nucleus in association with packaging proteins such as histones. The number of chromosomes is usually constant for a particular species and species with different numbers of chromosomes may have difficulty hybridising.
2. the structure that results from the packaging of DNA in higher plants and animals.

3. the bodies that contain the essential material of genetic coding.

- **chrysocentrus**
  with a golden centre.

- **cicatricose**
  describing spores marked with scars.
  Example Cicatricosisporites.

- **ciliata**
  (L. *cilium*, eyelid) with marginal hairs that form a fringe.

- **ciliate**
  1. fringed with very fine hair-like filaments.
  2. regarding or resembling the action of cilia.

- **ciliolate**
  marginally fringed with tiny hairs.

- **cilium**
  a small hair, often motile.

- **cincinnus**
  a form of definite flowering in which the successive axes arise alternately to the right and left of the preceding one, in distinction from the bostryx, in which the growth is all on one side; a *scorpioid cyme*.

- **cingulum / cingula / cingulate**
  a thick outer structure of a spore that projects at the equator, but does not extend over the distal or proximal face. Example Densosporites anulatus.

- **circinate**
  1. circular or ring-shaped. Applied to leaves that are rolled up on itself from the tip toward the bottom, like a shepherd's cane. (i.e. fern fronds and the leaves of the Sundew -Drosera).
  2. rolled into a coil with the apex at the centre.

- **circinate vernation**
  the coiled arrangement of young leaves of ferns & cycads.

- **circular-bordered pit**
  a round pit with a thickened margin.

- **circumaperturate**
  describing a pollen grain with equatorial apertures that are regularly arranged around a circular outline.

- **circumnutation**
  the continuous motion of some part of a plant, as the top of the stem or a tendril, in which it describes irregular ellipses or circles.

- **circumpolar lacuna / circumpolar lacunae**
  a lacuna in lophate pollen grain immediately adjacent to a polar lacuna that is not part of the apertural system.

- **circumscissile**

1. opening or divided by a horizontal circular line, applied to pods or seed capsules that open with a lid, like a box. The fruit in such cases is called pyxidium.
2. (L. *circum*, around *scindere*, to cut) opening splitting by a transverse fissure around the circumference, leaving an upper and lower half; said of certain seed pods or capsules.

- **cirrhous**

terminated by a tendril, being a projection of the midrib.

- **cirrose**

1. a leaf tipped with a tendril or, in mosses, with a very narrow or hair-like sinuous point.
2. resembling tendrils or coiling like them. Also cirrate, cirrous, cirrhous, cirrhose.

- **cirrus**

a tendril. A long thread-like part by which some plants climb. Also cirrhus.

- **cladistics**

a method of classification of organisms based on their common ancestry and exploring their relationships within the branching evolutionary family tree or cladogram. A groups of organisms with common ancestors and therefore sharing similar characteristics, is called a clade. Cladograms are graphical representations of the evolutionary divergence of species or related groups of species (clads) from common ancestors.

- **cladocarpous**

having the fruit at the end of short side branches.

- **cladode**

1. flattened leaf-like stem.
2. a stem that is leaf-like in appearance.

- **cladogram**

a dichotomous phylogenetic tree that branches repeatedly, suggesting a classification of organisms based on the sequence in which evolutionary branches arise. A nested diagram of synapormorphies indicating relations between groups, each point of branching represents divergence from a common ancestor.

- **cladoptosis**

the annual falling of leafy twigs instead of individual leaves. (An example of this process takes place in many trees of the Cypress family.)

- **clasping**

wholly or partially surrounding the stem.

- **classification of plants**

taxonomy; the systematic naming of plants devised by Carl von Linne.

- **classopollis**
an extinct genus of conifer of the Mesozoic era with distinct pollen grains.

- **clava / clavae / clavate**
a club-shaped element of the sexine/ectexine that is higher than 1μm, with diameter smaller than height and thicker at the apex than the base.

- **clavata**
(L. *clava*, a club) club-shaped; having the form of a club, growing gradually thicker toward the top, as certain parts of a plant.

- **claviform**
club-shaped, tapering upwards from a narrow base.

- **claw**
1. the long and narrow base of a sepal or petal (including *labellum*), as in pinks, wallflowers and Orchids.
2. a sharp tool usually having three tines, used to weed and cultivate.
3. (AS. *clawu*, a claw, hoof) the narrowed, stalklike base of some sepals or petals.

- **clay**
a type of Soil that contains 10% or more clay. A soil particle less than 0.002 mm in diameter.

- **clay loam**
soil containing 27% to 40% clay.

- **clay mineral**
a naturally occurring inorganic crystalline material with particles less than 0.002 mm in diameter.

- **clay pan**
soil rich in clay. Soil hard when dry, stiff when wet and it will separate abruptly from overlying soil.

- **clayey soil**
soil made up of fine particles that won't hold much air.

- **cleft**
(AS. *cleofan*, to cut) divided halfway down to the midrib or further or generally, any deep lobe or cut.

- **cleistogamous**
1. of flowers that do not open, but nevertheless set seed by self-pollination.
2. capable of pollination without the flower opening; self-pollinating.

- **cleistogamy**
1. when a plant has regular, fully developed flowers as well as flowers in which development is checked in the bud, but are still fertile and produce a lot of seed. These undeveloped flowers are inconspicuous, non-

opening and lack petals, scent or nectar. They have small anthers that contain a few pollen grains and a dwarfed pistil. They are always fertile while the more perfect flowers of the same plant are usually nearly or completely sterile. Cleistogamy is known to happen in about 60 groups belonging to many different families, including Dicotyledonous. The violet is a good example. See **viola**.

2. (Gr. *kleistos*, closed gamos, marriage) the condition of having flowers which never open and self-pollination occurs and the flowers are often small and inconspicuous.

- **climate**

the amount or variations of cold and heat, dryness and moisture, calm and wind etc. in a region or country. The combined result of all the meteorological phenomena of any region, as affecting its vegetable production.

- **climber**

a plant that clambers upward by attaching itself to other plants or objects (such as a trellis). Climbing plants are distinguished as stem climbers, which wind upward around an erect support and, as tendril climbers, which cling to nearby objects by slender, coiling tendrils. Other plants climb and cling by means of spines or bristles or by rootlets.

- **clinandrium**

a cavity, at the apex of a column in orchids, in which the anthers rest.

- **cloche**

a bell-shaped cover used to protect a plant from frost or cold. An individual cold frame.

- **clone**

1. group of individuals originally taken from a single specimen and maintained in cultivation by vegetative propagation. All clone specimens are exactly alike and identical to the original. Most hybrids are clonal in origin and are normally propagated vegetatively.

2. (Gr. *klon*, a twig) a group of plants all of whose members are directly descended from a single individual.

3. a plant produced by asexual reproduction, e.g. from cuttings.

- **closed**

root cap is distinct from the root proper.

- **clustered**

several vessel occur in a bundle.

- **clustered stomata**

stomata occur in groups.

- **clypeate**
  shield-shaped, almost the same as scutate.
- **coalescent**
  growing together. Uniting so as to form one body. Applied to the organic cohesion of similar parts.
- **coaperturate**
  describing permanent tetrads in which the apertures of neighbouring monads join. Examples Erica (Ericaceae), Periploca (Periplocaceae).
- **coarse sand**
  a soil particle that measures between .50 to 1.0 millimeters.
- **coccineus**
  scarlet.
- **coccolithophore**
  a unicellular planktonic organism of uncertain type (protozoan or algae?). Currently assigned the phylum Haptomonada.
- **coccoliths**
  minute calcium carbonate platelets secreted by coccolithophores which bear them as surface plates.
- **coccus**
  one of the separate divisions of a divided seedpod, which splits up into one-seeded cells.
- **cochlear**
  shell-shaped.
- **cochleate**
  shaped like a snail shell.
- **coenobium**
  a colony of a fixed number of cells.
- **coenocytic**
  multinucleate, formed by nuclear division without cross wall formation.
- **coenocytic**
  multinucleate, without cell cross walls, composed of multinucleate cells or segments.
- **cold frame**
  a bottomless box consisting of a wooden or metal frame with a glass or polyethylene top. It is placed on the ground over plants to protect them from cold or frost.
- **coleoptile**
  in monocots, a sheath the covers the plumule.
- **coleorhiza**
  in monocots, a sheath that covers the radicle.
- **collar**
  1. a ring placed loosely around the stem of a plant, especially tomatoes, to prevent cutworm damage.

2. (L. *collare*, a band or chain for the neck) region of junction between blade and leaf sheath of grasses.

- **collateral bundle**

contains a mass of phloem toward the exterior and a mass of xylem toward the interior.

- **collenchyma**

1. tissue composed of unevenly thickened cell walls; collenchyma cells are flexible and support young parts of the plant without hindering growth; collenchyma cells are composed of cellulose.
2. a layer of tissue right beneath the epidermis, having the cells thickened at the angles with a pad-like mass, which swells in water. It is found mainly in the young stems, petioles and leaf veins of many dicotyledonous plants.
3. tissues composed of cells with unevenly thickened primary cell walls that strengthen growing organs. Types are classified according to the arrangement of the wall thickenings and include.
4. (L. from Gr. *killa*, glue and *enchyma*, an infusion) living, supportive tissue with chloroplasts generally just beneath the surface consisting or more or less elongated cells usually thickened unevenly in a manner somewhat variable in different groups of plants.

- **colliculate**

covered with small, rounded or hill-like elevations (colliculae) e.g. surface of seeds.

- **colloid soil**

organic or inorganic tiny crystals that are small enough to be easily carried by air or water.

- **colluvium**

soil that is deposited at the base of a steep slope.

- **colonial**

(Fr. *colonial*, from L. *colonia*, a colony) usually used to describe cloning by vegetative reproduction, the seemingly separate plants having arisen from rhizomes, stolons or roots of a single or of neighbouring 'parent' plants.

- **colony**

aggregation or loose organization of similar cells (or organisms), which developed from a single original cell or organism, each capable of independence from the rest of the colony

- **coloratus**

coloured, usually with a reddish tint.

- **colpate**
(Gr. *kolpos* + *E-ate*, of pollen grains) having longitudinal germinal furrows in the exine.

- **colpodiporate**
synonym of diploporate.

- **colporoidate**
describing pollen grains with colpi and indistinct ora.

- **colpororate**
a compound aperture characterised by an ectoaperture, a shorter lolongate mesoaperture and a lalongate endoaperture. Example Sonc-hus.

- **colporus / colpori / colporate**
a compound aperture consisting of an ectocolpus with one or more endoapertures.

- **colpus / colpi / colpate**
an elongated, aperture with a length/breadth ratio greater than 2.

- **colpus equatorialis / colpi equatorials**
synonym of endocingulum.

- **colpus membrane**
the aperture membrane of a colpus.

- **colpus transversalis / colpi transversales**
synonym of endocolpus. See **endoaperture**.

- **columella / columellae / columellate**
a rod-like element of the sexine/ ectexine, either supporting a tectum or a caput. See **baculum, pilum.**

- **columellae layer**
synonym of infratectum. See **interstitium**.

- **column**
1. an organ formed by the union of filaments with one another, as in Malvacea or of stamens with the style and stigma, as in Orchids.
2. the plant structure that supports the lid of a pitcher.

- **columnar**
tall and cylindrical or tapering.

- **column-foot**
a basal extension of the column to which the *labellum* is attached, often flexibly.

- **column-wings**
paired protuberances or wing-like flanges on the column.

- **coma**
1. the leafy head of a tree (such as Palms) or a tuft of leaves or bracts terminating a stem (such as the leafy top of a Pineapple). The silky hairs at the end of some seeds (such as fireweed - Epilobium).

2. (L. *coma*; Gr. *kome*, hair) a tuft of soft hairs, as at the apices or bases of seeds. A bunch of branches. A terminal cluster of bracts on a flowering stem, as in pineapples.

- **commensal**

an organism that lives on or with another without harming it. See **parasite, symbiosis.**
2. organisms living in, on or with another, without any particular advantage or harm from the relationship.

- **commissure**

1. a joint, seam, suture or closure; the place where two bodies or parts of a body meet or unite. The face by which one carpel or mericarp attaches to another, as in Umbelliferae.
2. (L. *commissura*, a joining together) a place of joining or meeting, as where one carpel joins another in the Umbelli-ferae.
3. the slit or line of dehiscence in the laesura. See **laesura.**

- **comose**

(L. *comosus*, hairy, from *coma*, hair) having a tuft of hair.

- **compact**

pressing air out of the soil.

- **compaction**

in sedimentation: the formation of massive rock from loose sediment, normally brought about by the weight of the overlying sediments, in preservation organic preservation in three dimensions, for example in peat or clay.

- **companion cell**

a specialised cell of phloem, derived from the same parent cell as the closely associated sieve-tube element immediately adjacent to it; the nucleus of the companion cell supports both its own cell andthe cell of its associated the sieve-tube element.

- **companion cells**

a cell in the phloem that is connected to a sieve-tube member by numerous plasmodesmata.

- **companion planting/cropping**

a system of arranging plants in a way that a given species benefits from some characteristic of another. For example, it is thought that Tomatoes are protected from certain insects by the odour of Marigolds.

- **compensation point**

conditions where respiration is in balance with photosynthesis, with generally no net gas exchange, in algae usually related to habitat depth.

- **competition**
(L. *competitio*, an agreement, rivalry) involves the removal or reduction of some factor from the environment by a plant or group of plants that is sharing the same habitat. Competition can be by an individual or groups of plants of the same or different species. Factors that may be reduced include water, minerals, food and light.

- **complanate**
lying in one plane. Applied to leaves, especially of Mosses.

- **complete dimorphism**
leaves completely fertile or vegetative.

- **complex soil**
1. a pattern of soil.
2. a mixture of different kinds of soil whose areas are too small to be shown individually.

- **complex tissue**
tissue that consists of more than one cell type, e.g. phloem.

- **composite aperture**
synonym of compound aperture.

- **compost**
the decomposition of a mass of rotted organic matter.

- **compost activator**
a substance, usually bacterial, that is used to speed the breakdown of organic waste into compost. Also called activator or compost inoculant.

- **compost aerator**
an elongated, flanged T-shaped device used to create passages for air and moisture in a compost pile. Also created compost turner.

- **compost bin**
a container to hold decomposing garden and yard wastes, fertiliser, peat, lime and other substances during the composting process.

- **compost inoculant**
a substance that may contain bacteria, fungi, enzymes and hormones that is used to speed the composting process and to improve the quality of the compost. Also called activator or compost activator.

- **compost tea**
a liquid fertiliser made of compost dissolved in water.

- **compost thermometer**
a thermometer having a long probe used to measure the temperature in the center of the compost pile.

- **composter**
a container utilising air, moisture and solar heat to turn grass clippings, garden waste and

kitchen scraps into compost. Composters may consist of stacked trays or other chambers arranged to facilitate periodic turning of the mixture.

- **compound**
having multiple parts or branches, not simple.

- **compound aperture**
an aperture with two or more components that are situated in more than one wall layer.

- **compound leaf**
a leaf that is divided into two or more leaflets.

- **compound leaves**
leaves divided into two or more leaflets.

- **compressa**
flattened or pressed together.

- **compressed**
flattened.

- **compression**
a fossil preservation type in which a thin film of organic matter is preserved. Compare to impression.

- **conceptacle**
1. an organ or a cavity that encloses reproductive bodies in some ferns, fungi, mosses and algae.
2. cavity or chamber containing reproductive structures with opening to the surface for release, e.g. in Corallinales and Fucales.

- **concordant pattern**
a pattern in a tectate pollen grain in which the arrangement of the columellae is the same as that of the elements upon the tectum. Example Lilium (Liliaceae).

- **concrescence**
the union of cell walls by means of a cementing substance formed in process of growth so that they are inseparably grown together. Also called cementation.

- **concretions**
pellet-like grains from concentrations of compounds in soils that cement the soil grains together.

- **conditional lethal genes**
genes that are expressed in response to a specific environmental stimulus such as a specific chemical. When expressed, it is lethal to the plant.

- **conditioner**
a material added to a fertiliser that keeps it flowing free.

- **conduplicate**
folded once longitudinally down the middle.

- **cone**
see **strobilus**.

- **conifer**
a plant that produces cones such as a plant belonging to the family Coniferae, such as Pines, Cypresses, Junipers and Cedars.

- **conjugation**
special form of isogamy where the reproductive protoplasts of two organisms form a common tube and fuse together, e.g. in Zygnemataceae.

- **conjunctate**
with bacula or columellae which are branched proximally into two or more parts.

- **connate**
of plant parts congenitally united into a single structure.

- **connation**
fusion of like parts eg. sepal with sepal, contrasted with adnation.

- **consistence**
properties of soil material that determines its ability to change shape.

- **consolidate**
to increase the density and reduce pore space in soil.

- **conspecific**
belonging to the same species.

- **consumptive use**
1. the water used by plants.
2. vapour loss. Expressed as equivalent depth of free water per unit of time.

- **contact area**
area on the proximal face of a spore interpreted as having been formed in contact with the other members of the tetrad. Example Retusotriletes pythovii.

- **contact pesticide**
a chemical that injures or kills insects which come into contact with it and does not have to be ingested to be harmful.

- **contour**
an imaginary line connecting points of equal elevation on the soil's surface.

- **contour basin**
a basin made by a levee or border built on a contour.

- **contour ploughing**
the ploughman keeps to a level line at a right angle to the direction of the slope.

- **contour terrace**
leveling soil out on a slope at a right angle to the direction of the slope.

- **contractile root**
a root that contracts to pull the crown of the plant below the soil surface.

• **conus / coni**
cone-shaped elements on the surface of spores in which the height is less than two times the basal diameter and the apex is pointed, blunt or rounded. Example Lophotriletes mosaicus.

• **convolute**
rolled together or one part over another. Specifically applied to leaves or petals rolled up lengthwise in the bud.

• **copal**
a resinous substance exuded from some species of tropical trees and hardening in air into a glassy solid ranging in colour from red to yellow or brown.

• **copper**
an element used in various forms, including copper sulfate and Bordeaux mixture, to control fungal and bacterial diseases of plants. Also used in strips around tree trunks to act as a barrier to snails.

• **copropalynology**
the study of palynomorphs in coprolites or excrement.

• **cordate**
(L. *cor, cordis*, a heart) with a sinus and rounded lobes at the base, the overall outline usually ovate; often restricted to the base rather than to the outline of the entire organ; heart-shaped.

• **cordiform**
heart-shaped, in the form of two rounded lobes.

• **coriaceous**
(L. *coriaceus*, from *corium*, leather) leathery, tough.

• **cork**
1. (periderm) specifically the soft, low density bark of the cork oak tree, composed of cambium cells. Cork contains a natural wax, suberin, which protects the tree from water loss. The stems of many succulent plants harden and go brown with age and are often referred to as 'corky'.
2. a plant tissue composed of cells whose walls are impregnated with suberin and are non-living at maturity, cork is produced by the cork cambium.
3. cells produced by the cork cambium that have suberised cells walls and are dead at maturity.

• **cork cambium**
a narrow cylindrical sheath of meristematic cells that produces cork cells to replace the

epidermis during secondary growth (growth in width).

- **cork cambium /phellogen**
  a sheet-like meristem that produces cork.

- **corking**
  gradual ripening of a stem from the base, to produce a bark-like texture.

- **corm**
  a flattened fleshy underground stem, often acting as a dormant resting stage, capable of producing stems from the base and top and leaves and flower stems from the top. A corm in a rounded, thick underground stem that is modified into a mass of storage tissue. The foliage and flowers grow from buds on the stem. Corms are usually round and slightly flattened, not pointed like a true bulb. If you cut one in half, you won't see scale rings. (see **Bulb** to distinguish.) The tunics of corms are usually netted and fibrous. One or more growing points or eyes, grow from the top of the corm. At the bottom, roots grow from a basal plate. As the plant grows, the old corm dies and new corms, called cormels, form around it. If a new cormel is large, it may produce flowers the following year, but normally it takes two to three years. Crocuses, montbretia and gladioluses are typical corms.

- **cormel**
  a smaller, secondary corm produced by an old corm. A bulb let.

- **cormophyta**
  (Gr. *kormus*, the trunk of a tree with the boughs lopped off *phyton*, plant) in older classifications, a division comprising all plants that have a stem and root.

- **corniculate**
  horn-shaped (of an extremity).

- **cornute**
  having horns.

- **corolla**
  1. the collective group of petals which may occur separately or fused into a cup, tube or other structure.
  2. the usually conspicuous part of a flower, called the petals. They are contained within the calyx and immediately surround the stamens and pistil.
  3. (L. *corolla*, a little crown) the inner, usually coloured or otherwise differentiated, whorl or whorls of the perianth; the petals of a flower as a whole.
  4. the innermost whorl of petals in a flower.

- **corona**

  1. a crown-like appendage or series of united appendages on the inner side of a corolla. It often resembles an outgrowth of the *perianth*, as in the Daffodil or of the staminal circle, as in the milkweed.
  2. an equatorial or subequatorial extension of a spore, resembling a cingulum, but divided into fringe-like elements (fimbria). Example Reinschospora speciosa.

- **coronule**

  a little crown of a seed. The downy tuft on seeds.

- **corpus**

  1. the inner layers that divide anticlinally or periclinally.
  2. the body of a saccate pollen grain or camerate spore.

- **corrugate**

  synonym of rugulate.

- **cortex**

  1. a soft layer of tissue between the woody vascular tissue and epidermis or external layer such as bark. Hence: *adj.* Cortical.
  2. a primary tissue composed mainly of parenchyma cells, which extends between the epidermis and the vascular tissue .
  3. the cylinder of primary tissue, within the stem of a plant, which extends from the vascular tissue in the centre out to the bark or epidermis. It may function in photosynthesis, food storage, support and growth of the plant.
  4. ground tissue between the vascular bundle and epidermis.
  5. tissue lying between the epidermis layer and the central medulla tissue.

- **cortical**

  relating to the cortex.

- **corymb**

  1. a flat-topped or convex flower cluster. The outer or lower flowers open first and have the longest stems.
  2. a flat-topped or dome-shaped flower cluster with the outer flowers opening first, as in species of the group Achillea.
  3. (Gr. *korys*, a helmet) a racemose type of inflorescence in which the lower pedicels are successively elongated forming a more or less flat-topped inflorescence, the outer flowers opening before the inner.

- **corymbose**

  having flowers in corymbs.

- **costa**

  1. the midvein of a minor divsion of a fern leaf.
  2. a thickening of the nexine/ endexine bordering an

endoaperture or following the outline of an ectoaperture.

- **costa equatorialis / costae equatoriales**
  synonym of costa endocinguli.
- **costate**
  ribbed.
- **cotyledon**
  1. the first embryonic leaf or leaves in the seed. In some plants the cotyledons remain underground in the seedcoat.
  2. the embryo leaf or leaves of seed plants that usually stores or absorbs food in a young seedling; in angiosperms (flowering plants) the following distinction is made with respect to the cotyledons dicots have two cotyledons and monocots have one cotyledon.
  3. the first leaf, pair or whorl of leaves to show after germinating, a seed leaf.
  4. the first leaves of an embryo, may or may not resemble true leaves.
  5. (Gr. *kotyle*, a hollow or cavity) the first leaf or leaves of a seed plant, found in the embryo of the seed which may form the first photosynthetic leaves or may remain below ground.
  6. an embryonic leaf within or newly emerged from a seed.
- **cotyliform**
  rotate, but with an erect limb.
- **cover cells**
  sterile cells cut off from during the formation of tetrasporangia in Rhodophyta.
- **cover crop**
  a crop of nitrogen fixing plants that is planted with or in rotation with, edible crops to increase nitrogen, stabilise soil, attract beneficial insects or to be used as green manure.
- **crassitude**
  synonym of a thickening.
- **crateriform**
  goblet-shaped.
- **creek**
  (ME. *creke*, *crike*, from ON. - *kriki*, bend, concavity akin to ON. *krikr*, bend, bay) a natural stream of water normally smaller than and often tributary to a river.
- **creep soil**
  the downward, slow, irregular mass movement of sloping soil.
- **creeper**
  a plant that grows along the surface of the soil or any other surface such as a wall or fence, by sending out rootlets from the stem. (i.e. Ivy -Hedera, Virginia Creeper - Parthenocissus

quinquefolia, Trumpet Creeper - Campsis radicans).

- **crenate**
  1. toothed with shallow, rounded scallops, as the edges of a leaf. When the scallops have smaller ones upon them, the leaf is said to be doubly crenate.
  2. (L. *crena*, a notch) having a notched, indented or scalloped edge, as certain leaves.
  3. furnished with convex-sided teeth.

- **crenolate**
  with shallow-toothed edges.

- **crenulate**
  minutely crenate.

- **crescentic**
  describing a pollen grain in (polar view with a very thick exine in the medium of the intercolpium, gradually thinning towards the colpi. Example Valerianella.

- **crest**
  also called fasciation, a mutation that results when the growing point of a plant forms a long line, rather that a single point. In botanical terms, it is usually signified by *cristate*.

- **crispate**
  curled or rippled at the edge such as the leaves of Pittosporum eugenioides. Also crisped with wavy, curled or crumpled margins.

- **crista / cristae / cristate**
  a crest-like ornamentation element, taller than it is wide, characterised by a narrowly curved base and a sharp upper edge. Example Cristatisporites.

- **cristate**
  crest-shaped or crested. An abnormal form of growth resulting from lateral distortion of the growing point.

- **crocks**
  broken pieces of clay pot, used to cover drainage holes of pots in order to provide free drainage and air circulation to the root system and to prevent the medium (i.e. soil) from washing down the holes.

- **cross**
  1. to produce a hybrid plant by cross-fertilising individuals of different varieties or species.
  2. a hybrid produced by cross-fertilisation.

- **cross tetrad**
  synonym of decussate tetrad.

- **cross-fertilisation**
  the fertilisation of the ovules of one flower by the pollen of another, either between plants of the same species or between individuals of different species,

resulting in the production of a hybrid.

- **cross-section**
  see **transverse section**.

- **crosswall**
  end wall of a cell in a filament.

- **croton pattern**
  a characteristic type of ornamentation comprising rings of five or six (sometimes more) raised, often triangular, sexine elements arranged around a circular area, usually formed by capitate columellae (pila). Example Croton, Jatropha (Euphorbiaceae), Callitriche antartica (Callitrichaceae), Pimelea arenaria (Thymelaceae).

- **crown**
  1. the part of a tree above the first branching. The area from which new shoots arise or the point at which the roots meet the stem.
  2. the basal part at soil level where roots and stem join and from where new shoots are produced; the tops of the rootstocks.
  3. (L. *corona*, a crown, wreath) that part of a stem at or just below the surface of the ground, an inner appendage of a petal or the throat of a corolla; an appendage or extrusion standing between the corolla and stamens or on the corolla, an outgrowth of the staminal part or circle as in milkweeds.

- **crozier**
  the spirally coiled 'fiddlehead' of an immature fern leaf. The coiled developing leaf of a fern.

- **cruciate**
  having leaves or flowers in the shape of a cross with equal arms, as certain members of the Mustard family. Generally cross-shaped. In the Rhodophyta, the type of division where tetra sporangium divides in two planes at right angles to each other.

- **crumb structures**
  the porous granular structure in soil.

- **crust**
  a thin, dry, hard layer of soil that forms on the surface of many soils when they are exposed to excessive heat.

- **crustaceous**
  having a hard, brittle texture. Having a hard shell.

- **crustate**
  describing an aperture membrane that is thickly covered with coarse granules.

- **cryptoaperture/ cryptoaperturate**
  an endoaperture which is not

apparent in surface view, because there is no ectoaperture.

- **cryptogam**
1. lower plants reproducing by spores rather than seeds, e.g. ferns, mosses, fungi etc.
2. a plant, such as a fern, which reproduces by means of spores rather than seeds.
3. an old term for a non-seed plant, cryptogam means hidden wedding, i.e. as they produced no seeds their sex lives were unknown for a long time.

- **cryptopolar**
describing a spore in which the distal and proximal faces have dissimmilar sculpturing and lacks tetrad mark. Example Calobryum dentatum, Haplomitrium hookeri.

- **cryptospore**
1. non-marine alete sporomorph with well differentiated contact areas but without (haptotypic) features such as laesurae.
2. spores and other cells of ancient organisms of unknown affinity. Not strictly palynomorphs because the cell walls are destroyed by routine palynological processing. Cryptospores are generally small (10 - 20 μm) cylinders and spheres, smooth to ornamented. They are the primary indication of life during the Archean (3.5 - 2.5 GY) and Proterozoic (2.5 - 0.5 GY). However similar remains can be recovered by similar processes in Cenozoic sediments and in meteorites.

- **cryptostoma**
cavity open to surface that contains hairs or pariphyses, e.g. in Fucales.

- **cubical**
forming a cube or appearing similar to a cube.

- **cucullate**
hooded.

- **cullion**
an orchid.

- **culm**
1. the jointed and usually hollow stem of grasses. It is in most cases herbaceous but is woody in the bamboo and some other stout species. The term is also sometimes applied to the solid, jointless stems of sedges.
2. (L. *culmus*, a stalk, stem) the stalk or stem for such plants as grasses and sedges, usually jointed and hollow.

- **cultigen**
a cultivated plant or group of specific rank for which a wild ancestor isn't known. (i.e. Zea Mays).

- **cultivar**

a plant originated or selected artificially. Names of cultivars may be appended after the species name, (e.g. *Lithops bromfieldii* 'Sulphurea') sometimes following the letters *'cv'.* A contraction of 'cultivated variety' (abbreviated to cv). A group (or one among such a group) of cultivated plants clearly distinguished by one or more characteristics and which retains these characteristics when propagated. A distinct variety or race of plants that originated under cultivation and persists under cultivation. A cultivar may or may not be referable to a botanical species. Cultivars are given a name, usually distinguished by the use of single quotation marks, as in Zinnia elegans 'Tom Thumb'. Named varieties of hybrid Tea Roses and of garden Apples are examples of cultivars.

- **cultural requirements**

the conditions which provide optimal growth of a plant in cultivation.

- **cuneate**

1. wedge-shaped with straight sides converging at the base.
2. wedge shaped, triangular with acute angle downward
3. a (L. *cuneatus*, wedge-shaped, from *cuneus*, a wedge) narrowly triangular with the acute angle toward the base; wedge-shaped; tapering toward the point of attachment.

- **cuneiform**

wedge-shaped. In the form of a rounded isosceles triangle with the blunt end outermost.

- **cuneus / cunei**

structure formed by the termination of the inner wall layer at the endogerminal which gives rise to a shallow V-shaped atrium that points towards the centre of the grain. Example Pseudotrudopoilis.

- **cuniculus / cuniculi**

a space between the margin of the spore body and the inner face of the equatorially expanded spore wall, located at the equator.

- **cupulate**

synonym of foveolate.

- **cupule**

1. structures that surround one or more ovules or seeds, the cupule lobes may be free or united.
2. a cup-shaped *involucre*, occurring especially in the Oak, Beech, Chestnut and Hazel, formed by fused bracts.

- **cupuliform**
cupola-shaped. Shallowly concave with an entire or nearly so, margin.

- **curvatura / curvaturae**
a line in trilete spores, extending from the extremities of the ends of the radii of the laesura and thus delimiting the contact areas.
Examples, Laevigatisporites glabratus, Divisisporites divisus.

- **curvimurate**
describing pollen grains with curved muri.

- **curvospina**
with curved spines.

- **cusp**
1. a sharp, stiff point.
2. (L. *cuspis*, a point) rigid, sharp point, especially on a leaf.

- **cuspidate**
ending in a rigid point. An example is a leaf that is tipped with a sharp point or spine, such as the leaves of thistles. Tapering evenly to a hard point.

- **cut back**
to prune.

- **cut flowers**
flowers that are cut off the plant and used as decoration, usually indoors in a vase of water along with other flowers.

- **cut-and-come-again**
any plant that is cut or sheared after flowering and blooms again (i.e. Petunia, Pansy). Said of plants that continue to bloom when flowers are removed or cut.

- **cuticle**
1. a thin wax-like waterproof layer, composed of cutin, covering the epidermis. The outer coat of the seed.
2. an impermeable layer of cutin on the outer walls of epidermal cells.
3. a water repellent layer that coats the outer cell walls of the epidermis on aerial parts of plants, composed of cutin with a surface coating of wax.
4. structureless layer on outer surface of some plants.
5. (L. *cutis*, skin) a continuous layer of fatty substances covering over the outer surfaces of the epidermis of plants. It contains cutin and protects against water and gases.

- **cutin**
1. a waxy, water-impermeable complex fatty substance impregnating the cell walls and as a separate layer, the cuticle, on the outer surface of the epidermis.
2. a waxy substance which, together with cellulose, forms the

outer layer of the skin of many plants.

- **cutting**
a section of a plant that is removed and used for propagation. A cutting may consist of a whole or part of a stem (leafy or nonleafy), leaf, bulb or root. A root cutting consists of root only. Other cuttings have no roots at the time they are made and inserted. A propagation that consists of part of the crown of a plant or of its aboveground portion and roots is called a *division.*

- **cyanthium**
the inflorescence of a *Euphorbia.*

- **cyathiform**
cup-shaped.

- **cyathium**
1. a type of flower characteristic of Poinsettia.
2. (Gr. *kyathos*, cup) a type of inflorescence characteristic of some members of Euphorbiaceae, consisting of a cup-like involucre bearing unisexual flowers, staminate on its inner face, pistillate from the base.

- **cyathus**
a small conical or cup-shaped organ or cavity of a plant.

- **cycle**
a closed circle or whorl of leaves.

- **cyclic**
arranged in circles or whorls.

- **cyclospermous**
having the embryo coiled about the central albumen, as the seeds of Caryophyllaceae.

- **cymbiform**
boat-shaped. Concave, tapering at each end and keeled.

- **cyme**
1. an inflorescence with the oldest flower in the centre of the cluster.
2. flat-topped or dome-shaped flower head with the inner flowers opening first. (i.e. buttercup and anemone)
3. (Gr. *kyein*, to swell) a cluster of flowers in which each main and secondary stem bears a single flower, the bud on the main stem blooming first; determinate inflorescence in which each growing point ends in a flower.)
4. an inflorescence in which the flowers open in order from the terminal point of the primary axis to those of the higher axial orders in turn.

- **cymose**
1. bearing a cyme, resembling or pertaining to a cyme.

2. bearing a cyme or cymes.
3. in the manner of a cyme.

- **cystocarp**
the carposporophyte and surrounding envelop of haploid (1N) female gametophyte tissue (or pericarp) in the Rhodophyta.

- **cystocarpic**
bearing cystocarps, signaling a fertilised female gametophyte in the Rhodophyta.

- **cystolith**
(Gr. *kystis*, bladder; *lithos*, stone) a mass of calcium carbonate concretion, occasionally silica, formed on ingrowths of modified epidermal cell walls in some plants, esp. of the Acanthaceae family.

- **cytochrome**
an iron containing pigment that plays a major role in respiration.

- **cytokinesis**
1. the process of division of a whole cell, distinct from karyokinesis division of the cell nucleus.
2. process of cell division, as opposed to nuclear division

- **cytology**
the study of cells.

- **cytoplasm**
the contents of the cell excluding the cell nucleus.

- **dactyloid**
finger-like in form or arrangement.

- **daedaleous**
truncate, but raggedly so.

- **damping off**
a potentially fatal fungal disease, usually of seedlings or cuttings. A common name for a soil disease caused by a number of different organisms. It causes seeds to rot or seedlings to collapse. The best ways to prevent this is by planting in sterile soil, treating seeds with a fungicide or soaking the soil with a fungicide also avoid over-watering and planting warm-season vegetables in cold soil.

- **darlingtonia (g)**
a monotypic genus of pitcher plant native to the Western US, Cobra Lily.

- **dasyacanthus**
with dense spination.

- **dasyphyllous**
having woolly leaves. Having thickly set leaves.

- **day neutral plants**
plants that flower regardless of day length.

- **deadheading**
the removal of dead flowers or *flower heads*.

• **dealkalisation**
the removal of exchangeable sodium from the soil by a process of leaching or chemically treating the soil.

• **decayed organic matter**
decomposed plant or animal matter, used to improve the soil.

• **deciduous**
1. shed periodically or annually or at maturity. Often applied to the seasonal shedding of leaves.
2. falling off; in plants shedding leaves annually.
3. losing its leaves annually at the end of the growing season, semi-deciduous plants lose only some leaves.
4. (L. *deciduus*, that which falls down) falling after completion of the normal function.
5. periodically shed (often of decumbent (adj) lying along the ground.

• **declinate**
bent or curved downward or forward.

• **decompose**
to release chemicals from organic matter by means of bacterial action.

• **decomposer**
an organism (fungus, bacterium etc.) that breaks down and digests the remains of other organisms.

• **decompound**
divided into a number of compound divisions, such as a leaf divided into leaflets or a cluster of flowers divided into numerous branchlets.

• **decumbens**
lying on the substrate with an ascending apex.

• **decumbent**
reclining with the tips ascending. A plant that has its base lying on the ground and a stem that grows upward. A (L. *decumbere*, to lie down) trailing on the ground and rising at the tip, as some stems.

• **decurrent**
of a terminal spine extended down the leaf margin.

• **decurrent**
1. extending down the stem (i.e. a leaf with a base extending downward along the stem).
2. (L. *decurrere*, to run down) extending downward, applied usually to leaves in which the blade is apparently prolonged downward as two wings along the petiole or along the stem. Running down the stem, facing downwards.

• **decurved**
curved downwards.

- **decussate**
 arranged in pairs alternately crossing each other at right angles.

- **decussate branching**
 a type of branching in which the appendages are attached in pairs either alternately (alternate decussate) or at right angles (opposite decussate) to one another; the result are four ranks or rows of appendages.

- **decussate tetrad**
 a multiplanar tetrad of pollen grains or spores arranged in two pairs lying across one another, the pairs (dyads) more or less at right angles to each other. Example orophea (Annonaceae).

- **deep percolation**
 the downward movement of water into the soil beyond the reach of a plant's roots.

- **deep soil**
 soil that is deeper than 40 inches and continues until it becomes rock or another strongly contrasting material.

- **deflexed**
 curved or bent back down on itself or towards the plant surface. Bent backward or downward.

- **deflocculate**
 to break up the soil into individual particles.

- **defluent**
 running downwar, *decurrent.*

- **defluvium**
 a falling off, such as the bark of a tree, from disease.

- **defoil**
 to strip a plant of leaves.

- **defoliate**
 to deprive of foliage, to cut or pick the leaves of a plant, especially too early.

- **defoliation**
 the loss of leaves (for instance, from the attack of insects), specifically, the dropping of leaves in autumn.

- **defoliator**
 that which defoliates or strips of growth, specifically, in entomology, any insect that destroys the foliage of trees.

- **deforest**
 to deprive of forests, to cut down and clear away forests.

- **deforestation**
 the act of cutting down and clearing away forests of a region or stretch of land.

- **degradation**
 the change of one kind of soil to a more highly leached soil.

- **dehisc (v)**
 to shed pollen or seeds.

- **dehisce**
  to open (i.e. the seed capsules or pods of plants).
- **dehiscence**
  1. the bursting open of a pod or capsule to release the seeds or of an anther to release pollen.
  2.. (L. *dehiscere*, to gape) opening and shedding contents, said of stamens and fruits.
- **dehiscent**
  1. splitting or bursting open along defined lines eg. of a fruit.
  2. opening, such as the seedpod of a plant.
- **dehisces**
  vt., to burst or split open, as the seed capsules of plants.
- **dehydrate**
  to lose a high proportion of water content.
- **dehydration**
  the loss of water.
- **deliquescent**
  a tendency to absorb moisture and become liquid.
- **deltoid**
  1. triangular and attached to the stem at the wide end, such as a deltoid leaf.
  2. (Gr. *delta* and *eidos*, form) shaped like the Greek letter delta, triangular in outline.
  3. a triangular based pyramid.
- **demicolpus / demicolpi / demicolpate**
  an ectocolpus divided into two parts. Example, Amylotheca (Loranthaceae).
- **dendrochronology**
  the science of dating events and variations in environment in former periods by comparative study of growth rings in trees and aged wood.
- **dendroid**
  tree-like. Dendriform, arborescent; branching like a tree.
- **denitrification**
  the process by which bacterial action reduces nitrates in the soil to ammonia or free nitrogen that can escape into the air.
- **denshire**
  to improve land by burning piles of earth, turf and stubble and then spreading the ashes over the ground as compost. Also densher.
- **dentate**
  1. with sharp teeth perpendicular to the margin.
  2. toothed with sharp teeth directed outward as in a dentate leaf or having tooth-like projections, such as a dentate root.
  3. (L. *dens*, a tooth) toothed, with large saw-like teeth on the margin pointing outward, not forward.

- **denticle**
  . (L. *denticulus*, little tooth) a small tooth or toothlike projecting point.

- **denticulate**
  1. minutely dentate.
  2. having small teeth. Finely dentate.
  3. minutely dentate. Projecting in the form of a small tooth. Slightly toothed.

- **depressed**
  flattened from above.

- **derived trait**
  same as apomorphy, a derived character / trait is inferred to be a modified version of a more primitive condition of that character and therefore inferred to have arisen later in the evolution of the clade.

- **dermal tissue system**
  tissues derived from the protoderm or cork cambium that cover the surface of the plant body. The dermal tissues are complex (composed of several cell types).

- **desalinisation**
  the removal of salts from saline soil.

- **desert**
  a very dry area that receives less than 10 inches (25 cm) of precipitation each year.

- **desert soil**
  light-coloured surface soil that develops in warm- to cool-arid climates.

- **desilting area**
  an area used to remove sediment from flowing water.

- **desmid**
  green algae that have conjugation of nonflagellate, amoeboid gametes.

- **desorption**
  the removal of absorbed materials from surfaces.

- **dessicated**
  dried, wrinkled.

- **detailed soil map**
  a map showing the different classifications of soil in a designated area.

- **determinate growth**
  a type of growth in which the axis ceases growing, usually after the apical meristem differentiates into a reproductive organ, such as a flower or a cone.

- **dextrorse**
  twining spirally upward around an axis, as in the hop or morning glory.

- **di-**
  a prefix for two.

- **diadelphous**

1. having stamens united in two sets by their filaments. In pea and bean flowers, nine out of ten stamens are usually united while one is by itself.
2. (from *di-*, twice and Gr. *adelphos*, brother *-ous*) in two sets as applied to stamens when in two, usually unequal, sets.

- **diageotropic**

growing at right angles to the direction of gravity.

- **diageotropism**

a turning of the roots, branches or rhizomes in a direction at right angles to that of gravity.

- **diagnostic**

the investigative tools and techniques used in biological studies. For example, a particular set of diagnostic tools may be used to detect and identify microorganisms in food, water, soil or other samples.

- **diaheliotropic**

the turning of the upper surfaces of a plant's leaves toward the sunlight, pertaining to diaheliotropism.

- **diaheliotropism**

the tendency of a plant to turn the upper surfaces of its leaves toward the light.

- **dialycarpous**

bearing fruit composed of separate carpels. Also apocarpous.

- **dialypetalous**

plants that have the corolla of the flower divided into distinct petals, also polypetalous.

- **dialyphyllous**

made of separate leaves. Applied to a calyx formed from several sepals or a corolla formed by several petals.

- **dialysepalous**

having a calyx made of separate sepals, polysepalous.

- **diandrous**

a flower with two stamens.

- **diatom**

member of Bacillariophyceae.

- **diatomaceous earth**

a material made from the skeletal remains of diatoms (single-celled algae). It is used as an insecticide. The sharp edges of the diatoms wound soft-bodied insects like ants and aphids. It is ordinarily used as a barrier but can be applied to leaves. It is most effective in hot, dry weather.

- **diatomaceous earth**

siliceous geological deposits made up of diatom frustules.

- **diatoms**
  microscopic photosynthetic organisms with siliceous secondary plant walls termed 'frustules'. Mostly 20 - 100 μm and generally of two forms- pennate or centric.

- **diazinon**
  a man-made insecticide widely used against many soil pests, especially in lawns; however, it is very toxic to birds.

- **dibble**
  a pointed tool used to make holes in the ground for seeds and seedlings.

- **dichasial**
  pertaining to or resembling a dichasium.

- **dichasium**
  a flower cluster with two main stalks or branchlets within the cluster.

- **dichlamydeous**
  having both a calyx and a corolla.

- **dichogamous**
  exhibiting or characterised by dichogamy. Also dichogamic.

- **dichogamy**
  a provision in hermaphroditic flowers to ensure cross-fertilisation (and prevent self-fertilisation) by a difference in the time of maturity of the anthers and the stigma.

- **dichotomous**
  1. forking equally, divided into two parts e.g. double-headed plants.
  2. with branches forking into two more or less equal parts.
  3. (Gr. *dichotous*, a cutting in two) having or consisting of a pair or pairs, paired.

- **dichotomous branching**
  a type of branching in which the apical meristem divides into two more-or-less equal apices; by repetition of this type of branching in various planes distinctive shoot systems may be produced.

- **dichotomous key**
  a systematic procedure for determining the identity of an organism (or other object) by making a series of binary choices, according to the characteristics of the object, each of which leads to other choices or to the correct name.

- **dichotomy**
  a mode of branching by repeated forking, as is shown in some stems, the veining of some leaves etc.

- **diclinism**
  the state of having stamens or pistils in separate flowers. The state of being diclinous.

- **diclinous**
  having stamens or pistils in separate flowers.

- **dicoccous**
  formed of two one-seeded carpels.

- **dicolpate, dicolporate, diporate**
  describing pollen grains with two ectocolpi, two compound apertures or two pores. See **disulcate, monoaperturate, pantoaperturate, tricolpate, zonoaperturate.**

- **dicolporate**
  synonym of diploporate.

- **dicotyledon**
  1. a large group of flowering plants with two cotyledons that initially emerge from the seed. All *Cactaceae* are dicotyledons.
  2. a plant that produces a newly emerged seedling with two seed leaves before it produces mature leaves. Monocotyledons produce a newly emerged seedling with only one leaf.

- **dicotyledoneae**
  one of two major classes of plants, the other being Monocotyledonae. Dicotyledons have two seed leaves. The mature leaves branching with net-like veins and the flowers have five or sometimes, four parts. They include deciduous trees, broadleaf evergreens, flowering perennials, most vegetables and weeds.

- **dictyogen**
  a monocotyledon having net-veined leaves.

- **dictyogenous**
  having the character of a dictyogen having the general character of an endogen, but with netted leaf-veins.

- **dictyostele**
  1. a dissected siphonostele with two or more overlapping leaf bases.
  2. a dissected solenostele with each individual bundle a meristele.
  3. a siphonostele with numerous leaf gaps; superficially appears to be composed of vascular bundles.

- **didymous**
  twofold, twin, growing double, such as the fruits of umbelliferous plants or the tubers of some orchids.

- **didymous**
  twinned.

- **didynamous**
  having four stamens in two unequal pairs, such as most mints.

- **didynamy**
  the condition of being in two unequal pairs, such as stamens.

- **dieback**
  the death of tips or shoots due to damage or disease.

- **differentiate**
  become specialised for a particular function

- **differentiation**
  the progressive physiological and morphological changes in cells, tissues organs or the whole plant during development from meristem tissue to a mature or adult form. Differentiated tissues are usually more functionally specialised than undifferentiated ones.

- **diffuse**
  spreading widely and loosely, extended, dispersed, scattered.

- **diffuse growth**
  generalised growth, not localised at apex or base

- **diffuse porous**
  annual rings contain large vessel throughout.

- **dig**
  to break up and turn over piecemeal, as a portion of ground.

- **digitate**
  having deep radiating divisions, like fingers, arising from one point; especially applied to compound leaflets that grow from the top of the leaf stem.

- **dimerous**
  having two parts in each whorl of flowers.

- **dimidiate**
  having one part so much smaller than the other as to appear to be missing or almost so.

- **dimorphic**
  having two forms.

- **dimorphism**
  the occurrence of two distinct forms of flowers, leaves or other parts on the same plant or on plants of the same species, as in some Eucalyptus.

- **dimorphous**
  existing in two forms of leaf or flower or other part on the same plant or on plants of the same species.

- **dinoflagellates**
  unicellular aquatic organisms, motile and heterotrophic, parasitic or photosynthetic. Generally with two flagella, one within an encircling sulcus, the second (longitudinal) oriented posteriorly. Life cycles various, but often with motile thecate stage having a cellulose cell wall and a resting cyst (hypnozygote) with a chitinous or 'dinosporinous' wall forming inside the theca. Form of the cyst ranges from similar to the theca (proximate)

to bearing several trumpet-shaped processes (chorate). Theca and cyst often comprised of polygonal plates separated by sutures.

- **dioeciopolygamous**
polygamous with a tendency to dioeciousness or to the prevalence of flowers of one sex on individual plants, such as a plant with mostly female flowers.

- **dioecious**
1. have only female reproductive structures and male plants will have only male with male and female flowers on separate plants.
2. having unisexual reproductive structures confined to separate plants, i.e. female plants reproductive structures. compare with monoecious.
3. bearing male and female flowers on separate plants, such as in holly and willow. Also dioecian.
4. used to describe plant species in which male and female sex organs are borne on separate individuals.
5. (two houses) producing male and female gametes on different individuals or clone.
6. (Gr. *di*, two *oikos*, house) said of a kind of plant having unisexual flowers, the male and female flowers on different individual plants.
7. of plants forming both male and female flowers.

- **dionaea (g)**
a monotypic genus of carnivorous plant limited to a small area of the US East coastal plain and equipped with a hinged trap which snaps shut on its prey in response to stimulation of trigger hairs Venus Fly Trap

- **diorate**
synonym of diploporate.

- **dipetalous**
having two petals.

- **diphyllous**
having two leaves.

- **diploid**
having the basic two sets of chromosomes.

- **diploporate**
describing an ectocolpus with two endoapertures. Example Didymeles (Didymelaceae). See **endoaperture area.**

- **diplotegia**
a seed capsule that develops from the ovary in some plants such as the iris, in which the ovary in situated below the floral envelopes.

- **diploxylonoid**
describing bisaccate pollen grains in which the outline of the sacci in polar view is discontinuous with the outline of the corpus so that the grains seem to consist of three distinct, more or less oval parts. See **haploxylonoid, Haploxylon-type, Sylvestris-type.**

- **dipterous**
having two wing-like appendages.

- **disboscation**
the act of disforesting. The act of converting woodland into farmland.

- **disbud**
to remove buds or shoots, to remove unnecessary buds of a tree, vine or flowering plant to encourage production of high-quality flowers or fruits. Also called *pinching back* or *stopping*.

- **disc**
an elevated region of the *labellum* (of an Orchid), normally in the midlobe, distinguished by color and/or accessory features such as calli, keels or lamellae.

- **disc flower**
one of the multiple inner tubular florets on flower heads of *Compositae* (Sunflower family).

- **disc flowers**
the radially symmetrical flowers of the head in Compositae, as distinguished from the ligulate ray flowers.

- **discal**
disc-shaped.

- **discoid**
1. pertaining to the button-like center of flowers belonging to the Compositae family (i.e. daisy, sunflower etc.)
2. having only the central part and not the outer petal-like parts of the flower, as in Tansy.
3. (Gr. *diskos*, a disk) having the form of a disk, discoid flower, a compound flower not radiated but with tubular florets.

- **discoidal**
orbicular with a distinct rounded border.

- **discordant pattern**
a pattern in a tectate pollen grain in which the arrangement of the columellae is different from that of the elements on the tectum. Example Geranium (Geraniaceae).

- **dish garden**
a collection of plants, generally small ornamentals, grown together in one container.

- **disinfection**
the destruction of germs of in-

fectious diseases.

- **disjunct distribution**
  discontinuous distribution of a species.

- **disk**
  1. any flattened circular organ of a plant, such as the adhesive ends of the tendrils of the Virginia creeper.
  2. the button-like centre of the flowers of plants such as sunflowers and asters.
  3. a fleshy enlargement of the receptacle in some flowers.

- **disk flower**
  in the family Compositae, one of the miniature, tubular components of the central area of the flower head.

- **disparity**
  distinct in morphological characters, morphological variation; compare to diversity.

- **dispermous**
  having two seeds.

- **dispersal**
  distribution, usually of seed. To disperse to dispel or scatter, in plants dispersal refers to mechanism of dispersing reproductive propagules like seeds or pollen

- **dispersal unit**
  the morphological unit in which mature pollen grains or spores are shed, which may range from individuals (monads), to pairs (dyads), groups of four (tetrads) or groups of more than four (polyads). Larger, indeterminate numbers of pollen grains or spores may also be dispersed as pollinia or massulae.

- **dispersion of soil**
  the breaking up of soil into very fine particles and its suspension in water.

- **dissected**
  deeply divided or cut into parts. Deeply divided into many narrow segments.

- **dissections**
  rounded to elongated cavities in a cingulum or zona. Example Vallatisporites ciliaris. Synonym of vacuoles.

- **dissepiment**
  a partition within an organ of a plant (i.e. the membrane that separates sections of the orange and other citrus fruits).

- **dissepimental**
  pertaining to or of the nature of a dissepiment.

- **dissilient**
  bursting open with some force, as the pod or capsule of Impatiens.

- **distal**
  1. towards the unattached or developing end of an organ.
  2. (L. *distare*, to stand apart) farthest away from the point of attachment or origin.
  3. a common descriptive term used in contrast to proximal, applied in palynology to features on the surface that face outward in the tetrad stage.

- **distal face**
  that part of a palynomorph that faces outwards the centre of the tetrad, between equator and distal pole.

- **distal pole**
  the centre of the surface of the distal face.

- **distichous**
  1. arranged in two vertical rows on opposite sides of an axis.
  2. with leaves in two rows.
  3. (Gr. *distichos*, having two rows) two-ranked, in the case of plants with alternate leaves, the arrangement is such that 1st is directly below the 3rd.

- **disulcate**
  describing pollen grains with sulci arranged in pairs. Two types of disulcate pollen are distinguished: equatorial disulcate, with opposing, eaquatorially arranged sulci and distal disulcate with paired sulci lying parallel to the long axis of the pollen grain on the distal face. Examples Metroxylon salomonense (Palmae) (equatorial disulcate), Chamaerops humilis (Palmae) (distal disulcate). See **dicolpate, dicolporate, diporate, geminicolpate.**

- **diurnal**
  having daily periodic cycle, e.g. endogenous circadian rhythm.

- **divaricate**
  spreading far apart. vt. (L. *divaricare*, to spread apart) to branch or spread widely apart. Spreading widely divergent

- **divarication**
  the act of branching off or diverging, separation into branches.

- **divergent**
  (L. *divergere*, to bend away) separated from one another, having tips further apart than the bases

- **diversity**
  a terms used to describe number of taxa (species, genera etc.) and their relative abundance, also species richness. compare to disparity.

- **diverticulate**
  (L. *divertere*, to turn aside) having short offshoots approximately at right angles to axis.

- **divided**

1. separated to the base or midrib (i.e. leaves of palms or clover).
2. (L. *dividere*, to divide) referring to the blade of an appendage when it is cut into distinct divisions to or almost to, the midvein.

- **division**

a method of propagation by which a plant clump is lifted and divided into separate pieces, which include roots and a *growing point*, during dormancy. The pieces are immediately replanted.

- **DNA**

deoxyribonucleic acid- the genetic material, contained in the chromosomes of all cellular organisms and some viruses eg. potato spindle tuber virus. An extremely long double-helical linear polymer in which the gene sequences are encoded as the order in which four types of nucleotide bases occur, each paired with a complementary base. Groups of three base pairs in a structural gene specify the code for the 20 types of amino acids (and one or two rare ones such as selenocysteine) that make proteins.

Other sequences encode control functions such as start or stop (transcribing a sequence) signals. Much of the DNA in many organisms consists of apparently non-coding or repetitive sequences of unknown function.

- **DNA (deoxyribonucleic acid)**

this molecule stores the genetic information of all organisms. DNA is composed of four building blocks called nucleotides, represented by A, C, G and T.

- **DNA sequence**

a sequence of nucleotides that can be from any DNA region, such as a gene or chromosome.

- **dodecamerous**

having the whorls of the flower (petals, sepals etc.) in twelves. Also written 12- merous.

- **dolabriform**

having the form of an axhead. Axe-shaped, in the shape of an axehead.

- **dolomitic limestone**

a fertiliser that contains magnesium.

- **dormancy**

the state of a plant when growth has ceased for the year and other activities in whole plants have slowed down, usually during winter.

- **dormant**

in a temporary state of suspended activity. Not in active growth, but protected (e.g. bud scales) from environmental stresses.

- **dormant spray**
  any type of pesticide applied to dormant, usually leafless, plants to control a number of insects and disease.

- **dorsal**
  1. on or attached to the back of an organ. Pertaining to or attached to the back of a plant or plant part, facing away from or situated outside the axis of a plant.
  2. (L. *dorsum*, the back) pertaining to the back, the surface turned away from the axis.

- **dorsifixed**
  attached at the back of a plant or plant part, such as the anthers of lilies.

- **dorsiventral**
  flattened and having a definite front and back surface, such as a blade of grass.

- **double**
  this is used to describe flowers which have more than one row of petals. This may be because the stamens and sometimes the pistils are transformed into petals or sometimes the petals split to form several more. Semi-double flowers are those which have only partially changed. Completely double flowers have lost their reproductive organs and are therefore unable to produce seeds. These flowers usually stay fresh longer than *single* flowers when cut. Double flowers are seen in the wild once in a while, but are mostly developed varieties.

- **double fertilisation**
  in flowering plants, the more or less simultaneous union of one sperm and one egg to form a zygote (N=2) and another sperm with two polar nuclei to form triploid (N=3) endosperm in the ovule; in the Gnetales, the fusion of two sperm with two eggs to produce two zygotes, only one of which will mature into an embryo.

- **double integument**
  the two outermost layers of an ovule in angiosperms, one of which will differentiate into the seed coat. See **integument.**

- **doubly serrate**
  alternating large and small teeth.

- **down**
  (ME. *down, downe*, down; probably of Scandinavian origin) fine, soft feathers, soft, fine hair.

- **downy**

  1. covered with soft hairs or down.
  2. covered with short, fine hairs.

- **drain**

  to remove water from the soil by artificial means, such as drainage ditches, buried perforated plastic pipes or a gravel sump. Also to improve heavy clay soils with sand or organic matter.

- **drainage**

  1. the removal of excessive water from the ground.
  2. water moving down in the soil.
  3. a system of conduits, channels or passages by means of which something is drained.

- **drain-well**

  a pit sunk through an impenetrable layer of earth or stone to a porous layer, to draw off water.

- **drepaniform**

  sickle-shaped.

- **dried blood**

  an organic fertiliser that leaves an acid reaction.

- **dried flowers**

  blossoms that have been dehydrated but have retained a natural shape and colour.

- **drift**

  soil material deposited by geological process in one place after having been removed from another place.

- **drill**

  1. a machine used for planting seeds, as of grasses, wheat, oats, corn etc. by dropping them in rows and covering them with earth.
  2. a row of seeds deposited in the earth; the trench or channel in which the seeds are deposited.
  3. to sow seeds in rows (i.e. the field was drilled, not sown *broadcast*).

- **drip irrigation**

  a system of watering by which moisture running through a porous hose is slowly released through tiny holes or emitters, to the root zone of plants.

- **drip tip**

  a sharp elongated point on which atmospheric water accumulates from dews, mists and rain until the water drop is large enough to fall off the tip, to the ground around the roots of the plant. Drip tips are common on the leaves of tropical trees in areas of high rainfall and the spines of some *Cactaceae* also function in this way.

- **drosera (g)**
an extremely widespread genus of over 100 species found in climates from tropical to boreal, Sundew

- **drosophyllum (g)**
a monotypic genus of carnivorous plant found on both sides of the straits of Gibraltar, it is a passive trap) resembling Drosera, Portuguese Sundew

- **drought-tolerant plants**
plants that can survive periods of time with little or no water.

- **drowning**
a surplus of water that drowns plant roots by eliminating oxygen from the soil.

- **drupaceous**
pertaining to or producing drupes.

- **drupe**
1. a stone fruit having a hard, nut-like inner part surrounded by a fleshy or fibrous outer layer (i.e. cherry, peach, olive, walnut); a small fruit with a single seed.
2. (Gr. *dryppa*, an overripe olive) a fleshy or pulpy fruit with the inner portion of the pericarp hard or stony and enclosing the seed; usually 1-locular and 1-seeded, sometimes more than 1-locular and more than 1-seeded.

- **drupelet**
a small drupe, especially as applied to the individual components of aggregate fruits such as raspberries and blackberries, which consist of numerous seeds, each surrounded by its own fleshy layer. Also drupel.

- **druse**
compound calcium oxalate crystal of more or less spherical shape, in which the many component crystals protrude from the surface giving a star-shaped appearance

- **druse crystals**
granular.

- **druxy**
pertaining to a tree or timber, having decayed spots or streaks in the heartwood.

- **dry sand**
a sandy deposit with low water-holding capacity.

- **duff**
the partly decomposed organic surface layer of forested soil.

- **dune**
a mound of loose sand piled up by wind.

- **duplibaculate**
synonym of duplicolumellate.

- **duplicolumellate**
  with columellae in two rows under each murus. Example Polygonum persicaria (Polygon-aceae).
  See **pluricolumellate, simplicolumellate.**
- **dust mulch**
  a dry, loose surface layer of cultivated soil, which reduces water loss from the underlying soil.
- **dust/dusting**
  method of pesticide control in which a dry substance is applied by spraying.
- **duster**
  a device consisting of a bin, a wand with a nozzle and a crank mechanism used for applying dust or powder to plants.
- **dwarf fruit trees**
  a small fruit tree reaching a height at maturity of 4 or 5 feet and bearing early, but normal fruit.
- **dwarf plants**
  plants of genetically small size or those dwarfed by grafting (i.e. dwarf apple trees) or by pruning and restricted root growth (i.e. bonsai).
- **dwarf shrubs**
  shrubs that are selected or hybridised that are typically shorter and more compact than the species.
- **dyad**
  general term for two microspores (pollen grains or spores) united as a dispersal unit. See **dispersal unit.**
- **dyeing flowers**
  blossoms capable of producing dyes.
- **eared**
  having ears or awns, such as in corn or wheat.
- **ebracteate**
  without bracts.
- **ebracteolate**
  without small bracts.
- **ecalcarate**
  having no spur.
- **ecarinate**
  without a carina or keel.
- **ecaudate**
  without a tail.
- **echinate**
  prickly, covered with stiff bristles.
- **echinocarpa**
  with spiny fruit. Also acanthocarpa.
- **echinolophate**
  describing a lophate pollen grain with echinate ridges. See **fenestrate, psilolophate.**

• **ecology**
1. the study of the interrelationships of plants, animals and their environment.
2. (Gr. *oikos*, house and *-logy* Gr. *-logia*, from *legein*, to speak) branch of science concerned with the interrelationships of organisms and their environments esp. as manifested by natural cycles and rhythms, community development and structure, interaction between different kinds of organisms, geographic distributions and population alteration.

• **ectexine**
the outer part of the exine, which stains positively with basic fuchsin in optical microscopy and has higher electron density in conventionally prepared TEM sections. Orthographical variant- ektexine. See **sexine, endexine.**

• **ecto-**
a prefix for outer.

• **ectoaperture**
an aperture in the outer layer of the sporoderm. Examples ectocolpus, ectopore.
See **endoaperture, mesoaperture.**

• **ectocarpin**
sex hormone produced by female gametes of Ectocarpus that causes male gametes to cluster near source of the hormone.

• **ectointine**
synonym of exintine.

• **ectomycorrhizae**
fungal mycelia are external.

• **ectoparasite**
parasite living outside its host

• **ectophloic**
having phloem only on the outer side of the stele; compare to amphiophloic.

• **edaphic**
relating to or determined by, conditions of the soil.

• **eddish**
stubble, a stubble field. Also called earsh and ersh.

• **edging**
1. any nonbiodegradable material ordinarily plastic or metal, installed in the ground to separate flower beds from lawn or other landscape areas.
2. a border, a skirting, specifically, a row of plants set along the border of a flower bed.

• **efflorescence**
the time or state of flowering. Anthesis.

• **efflorescent**
blooming or flowering.

- **ejectosome**
  type of ejectile organelle found in cryptomonads.
- **ektannulus / ektannuli**
  an abrupt thickening of the outer wall layer in the region of the exogerminal. Example: Krutzschipollis.
- **ektexine**
  orthographic variant of ectexine.
- **ekto-**
  orthographical variant of the prefix ecto-.
- **elaioplast**
  contains oil droplets, usually found in fruits or seeds.
- **elater**
  1. one of four elongate appendages on the spores, as in *Equisetum.*
  2. a spirally twisted, filamentous band attached to certain spores and some fossil pollen grains. Example Equisetum (Equisetaceae).
- **electric hotbed**
  a cold frame used for extending the growing season. They are usually heated by an electrical mat or heating element.
- **ellipsoid**
  a compressed or flattened sphere.
- **ellipsoidal**
  in the form of an ellipse.
- **elliptic**
  1. broadening at or about the center and narrowing equally toward each end.
  2. (Gr. *elleipsis*, a falling short, defect, ellipse) an outline that is oval, narrowed to rounded at the ends and widest at about the middle (as the outline of a football), ellipsoid, a solid with an elliptical outline.
- **elliptic, elliptical**
  oblong with regular flattened ends.
- **elongate**
  lengthened.
- **eluviation**
  the movement of moisture and nutrients from one place to another in the soil.
- **emarginate**
  1. having a shallow notch at the tip, such as on a leaf.
  2. vt. (L. *emarginare*, to deprive of the edge) said of leaves, sepals or petals and other structures that are notched at the apex.
  3. a rounded end terminated by a notch, as if a piece has been removed. The leaf margin is coloured distinctly from the lamina.

- **emargination**
  a notching of the edge, such as along the edge of a leaf.

- **embryo**
  1. the new plant developing from the fertilised egg cell (zygote) in the fertilised seed.
  2. the developing immature plant contained in the seed of flowering plants prior to germination and growth.
  3. a young plant present in the seed before germination.
  4. the immature living cell-bundle formed by fertilisation.

- **embryo sac**
  the eight-celled megagametophyte of flowering plants.

- **embryophyte**
  the group of all organisms that retain the zygote on the parent to form an embryo.

- **embryophyte spores**
  a wide variety of microscopic unicellular reproductive cells developing into gametophytes. Examples include liverworts, club mosses, horse tails, ferns and many extinct forms. Typically with a triangular or oval outline and various scultpure.

- **embryotega**
  a small circular thickening near the seed scar on the seeds of some plants, such as members of the spiderwort family.

- **emerge**
  vt. (L. *emergere*, to rise up, rise out) to rise out of a fluid or other covering.

- **emergent**
  (ME. Fr. L. *emergent-*, emergens, pres. part. of *emergere*, to emerge - more at emerge) any of various plants (as a cattail) rooted in shallow water and having most of the vegetative growth above the water.

- **emersed**
  1. rising above the water, such as the leaves of many water plants.
  2. standing out of or rising above a surface as an aquatic plant with flower stalk emersed.

- **emersed plants**
  plants growing with their roots and a portion of the shoot below the water and the remainder of the shoot above the surface of the water.

- **enantioblastic**
  pertaining to a plant embryo originating at the end directly opposite the hilum.

- **enation**
  1. a non-vascularised, epidermal outgrowth found in some early land plants.
  2. an outgrowth from the surface of a leaf or other plant part.

3. (L. *enasci*, to spring up) an abnormal growth of an organ or of an excresence upon any part of a plant.

- **endannulus / endannuli**
  an annulus formed by the endexine of a pollen grain. Example- Atlantopollis.

- **endarch**
  a type of xylem maturation in which protoxylem is internal to metaxylem and development prodeeds centrifugally (from the inside out); for comparison see **exarch and mesarch.**

- **endemic**
  1. originating from and confined to a specific, usually small, geographic area.
  2. native to, naturally restricted to or found only in a particular area or region.
  3. (Gr. *endemos*, native, belonging to a people) a plant that is native to a particular country or region, not introduced or naturalised.
  4. restricted to a certain geographical location.

- **endexine**
  the inner part of the exine which remains relatively unstained with basic fuchsin in optical microscopy and has a lower electron density in conventionally prepared TEM sections. Synonym of nexine 2. (Erdtman, 1960b). See **ectexine.**

- **endintine**
  the inner, cellulosic zone of the intine which is adjacent to the cytoplasm and in fluorescence microscopy stains positively with PAS calcofluor.

- **endo-**
  a prefix for inner.

- **endoaperture**
  an aperture in the inner layer of the sporoderm, often the inner aperture of a compound aperture. See **ectoaperture, mesoaperture.**

- **endoaperture area**
  the region of the nexine of a single compound aperture which encloses a number of endoapertures.

- **endocarp**
  in fruits whose *pericarp* consists of dissimilar layers, the innermost layer. It may be hard and stony (i.e. plum or peach pit), membranous (i.e. apple core) or fleshy (i.e. orange pulp).
  (Gr. *endo*, within *karpos*, fruit) the inter layer of the wall of a matured ovary. When its texture differs from the outer wall, it may be hard and stony, membranous or fleshy. The innermost layer of the pericarp.

- **endocingulum / endocingula / endocingulate**
  a ring-shaped endoaperture continuous around a pollen grain and lying in the equatorial plane. Example Polygonum convolvulus (Polygonaceae).
  Synonym of colpus equatorialis and zonorate.

- **endocrack**
  an irregular groove occurring in the inner surface of the nexine/endexine and readily apparent in acetolysed pollen. Example Anemone nemorosa (Ranunculaceae).

- **endocytic**
  living inside a cell.

- **endodermis**
  characterised by a suberized Casparian strip which regulates transport of materials into the vascular bundles of most roots and some leaves and stems.

- **endogamous**
  self-pollinating.

- **endogen**
  a plant that develops from or on the inside, as most monocots.

- **endogenous**
  1. of or pertaining to endogens (endogenous plants).
  2. proceeding from within (endogenous growth).

- **endogerminal**
  describing an aperture in the inner wall layer.

- **endointine**
  orthographic variant of endintine.

- **endomycorrhizae**
  fungal mycelia are internal.

- **endophytic**
  living within tissues of host plant or alga.

- **endoplica / endoplicae**
  a fold or swelling of the inner wall layer.

- **endosculpture**
  sculpturing occurring on the inner surface of the nexine/endexine.

- **endosperm**
  1. mass of nutrient tissue formed within the embryo sac.
  2. the triploid (N=3) product of double fertilisation in angiosperms; during seed maturation the endosperm will develop into a storage tissue that will provide nutrients to the seedling as it emerges (in monocots) or that will be digested and stored by the cotyledons before germination (in dicots).
  3. a triploid nutritive tissue that develops in the ovule, may be absorbed by the before the seed matures.

4. nutrients enclosed in the seed to sustain early growth of the embryo.

- **endospore**
the innermost layer of a spore wall.

- **endosporic**
gametophyte develops within the spores wall, compare with exosporic.

- **endosporium**
synonym of intine.

- **endospory**
a condition in which the gametophyte develops within the spore wall, rather than externally.

- **endostome**
the opening in the inner covering of the ovule.

- **endosulfan**
a broad spectrum, synthetic insecticide used to control many plant pests, including whiteflies and thrips. Common trade name is Thiodan.

- **endotesta**
the delicate, innermost layer of the integument in some seeds, compare to sarcotesta and sclerotesta.

- **endozoic**
living with tissues of animal host.

- **ensiform**
1. sword-shaped; long, narrow, flat and tapering to a point, as the leaves of many species of iris.
2. (F. *ensiforme*, Fr. L. *ensis* sword + F. *forme*, form) having sharp edges and tapering to a slender point having a shape suggesting a sword.
3. sword-shaped, lingulate, tapering to a fine point.

- **entire**
1. with a continuous margin without teeth, lobes or indentations.
2. undivided and without teeth.
3. (L. *integer*, whole, untouched, undiminished) having a margin devoid of any indentations, teeth or lobes.
4. totally free from division or interruption.

- **entire margin (leaves)**
margin forming a smooth line or arc without noticeable serrations, note that lobed leaves can also have entire margins. Compare to lobed margin and toothed margin.

- **entomophilous**
1. applied to flowers whose structure ordinarily requires insects to pollinate them.
2. pollinated by insects.
3. (Gr. *entomon*, insect; *philein*, to love) pollination by insects.

- **environmental palynology**
the use of palynomorphs, their identification, distribution and abundance to determine past changes in the biota, climate or surficial geology of an area. Includes Quaternary palynology and Archaeological palynology . Exclusive of stratigraphic palynology, but the study of sedimentary sequences often includes both stratigraphic and environmental palynology.

- **enzyme**
a protein molecule that catalyzes biochemical reactions in living systems, including those found in metabolic pathways. As a catalyst, an enzyme facilitates reactions without being consumed by them. Examples include the stomach enzyme pepsin, which digests food and cellulase, which digests the polymeric sugar cellulose into smaller units. One of a number of chemicals secreted by a carnivorous plant to assist in the digestion of captured prey.

- **epappose**
having no tuft of bristles, a feature often found on seeds that are scattered by the wind, such as the seeds of a dandelion.

- **ephemeral**
1. a plant that completes its life-cycle in a very short time.
2. (Gr. *ephemeros*, lasting for a day) referring to an organ living a very short time, usually a day or less; lasting a very short time.

- **epicalyx**
a calyx-like involucre of bracts below the true calyx, as in the mallow or of sepals, as in the potentilla.

- **epicarp**
the outer skin of fruits.

- **epichilium**
the terminal part of the lip of some orchids. Also epichil or epichile.

- **epicone**
part of dinoflagellate cell anterior to the cingulum.

- **epicotyl**
1. the main axis of the seed embryo just above the cotyledons and continuing into the developing stem of the seedling.
2. the part of a plant embryo or seedling above the cotyledons.

- **epicotyledonary**
pertaining to the epicotyl. Situated above the seed leaves (cotyledons).

- **epidermis**

1. the thin outermost protective cell layer, usually taken to include the waxy water-impermeable material (cuticle) overlying the living cells.
2. the exterior tissue, usually on cell thick, of leaves and young stems and roots.
3. the outer, protective tissue of a plant. In succulents, this may bear protective features such as thick, fine hair or bloom.
4. a complex tissue that is usually a single cell layer thick and composed of the following cell types.

- **epigeal**

1. growing with one or more leaves above the ground. Germinating with cotyledons above the ground.
2. a plant that grows above the surface of the soil, as most plants in the garden do.
3. a cotyledon that is forced above the ground in the process of germination, as in beans, morning glories and others.
4. cotyledons that appear above ground. Also epigeous, epigean, epigaean or epigeic.
5. growing with one or more leaves above the ground - the usual habit of plants.

- **epigenous**

1. growing on the surface of a leaf or other plant organ (i.e. fungi grow on the surface of leaves, such as mildew).
2. referring to the calyx, corolla and stamens that appear to be growing from the top of a ovary, but are actually fused to it, as in the cranberry.
3. sepals, petals and stamens attached to the top of the ovary.
4. (Gr. *epi*, upon *gyne*, woman) growing upon the top of the ovary or seeming to do so, as petals, sepals, and stamens.

- **epigyny**

the condition of being epigynous.

- **epilithic**

living on rock or stone.

- **epinastic**

growing faster on one side of a leaf or stem than on the other - the means by which many active traps move.

- **epipetalous**

borne on the petals or the corolla of a flower; applied to stamens and to plants whose stamens are attached to the corolla. (Gr. *epi*, upon *petalon*, leaf) having stamens inserted on petals.

- **epipetiolar**

with branches arising from buds on the petiole.

- **epiphyte**
  1. a plant (or other organism) that grows on another for support, without drawing any nutrients from it. Hence Adj. *Epiphytic.*
  2. a plant that grows on another (such as on a tree) without being parasitic and obtains nutrients from decaying leaves and other debris and moisture from the air without rooting into the soil. These plants do not obtain nutrients directly from the host plant as do *parasites* and *saprophytes.* Many Orchids, Bromeliads, Ferns and some Cacti are epiphytes.
  3. a plant that grows on another plant. No parasitism is involved. The epiphyte must be rooted on the surface of the host. Vines rooted in the soil and climbing up on another plant are not epiphytes.

- **epiphytic**
  living on a plant. Or growing on other plants (usually trees,) without deriving or contributing nutritional benefit from them.

- **episepalous**
  growing on or fused with a sepal, applied to stamens.

- **episperm**
  the hard external covering of a seed.

- **epispore**
  the outer coat or membrane of a spore.

- **epistomatic**
  stomata in the upper epidermis only.

- **epitheca**
  the epivalve and epicingulum of a diatom frustule or cell wall.

- **epithet**
  a descriptive adjective used as a noun forming part of the binomial botanical name of a species. e.g. gracilis = slender.

- **epixylous**
  growing on wood.

- **epizoic**
  living on an animal.

- **epruinose**
  having no *bloom* or powdery coating upon the surface of the leaf, stem or other plant part, not *pruinose.*

- **equator**
  the dividing line between the distal and proximal faces of a pollen grain or spore. See **equatorial diameter, equatorial plane.**

- **equatorial axis**
  often misappropriately used as a synonym of equatorial diameter.

- **equatorial bridge**
  synonym of bridge.
- **equatorial diameter**
  a line, lying in the equatorial plane, perpendicular to the polar axis and passing through it.
- **equatorial lacuna / equatorial lacunae**
  a lacuna situated on the equator in the mesocolpial region of a lophate pollen grain. example Tragopogon pratensis (Compositae). See **lacuna.**
- **equatorial limb**
  synonym of equatorial outline.
- **equatorial outline**
  general description of the equator when a pollen grain is seen in polar view. See **amb.**
- **equatorial plane**
  the plane perpendicular to the polar axis and lying midway between the poles.
- **equatorial ridge**
  a ridge lying along the equator in the mesocolpial region, usually applied to lophate pollen grains. Example Cichorium intybus (Compositae).
- **equatorial view**
  the view of a pollen grain or spore where the equatorial plane is directed towards the observer.
- **equiaxe / equiaxal**
  pollen grains with a polar axis equal to the equatorial diameter. See **breviaxe, longiaxe.**
- **equitant**
  leaves that overlap each other to form a fan (i.e. as many varieties of Iris). Or (L. *equitare*, to ride) overlapping, said of leaves whose bases overlap the leaves within or above them, as in the Iris.
- **erect**
  synonym of prolate.
- **erectocentrus**
  with an erect central spine.
- **ericaceous**
  applied to acid potting mix with a pH of 6.5 or less. It is usually used for plants that cannot tolerate lime.
- **ericetal**
  composed of or containing heaths.
- **erinaceae**
  with a hedgehog-like appearance.
- **eroded**
  irregularly bounded, as if gnawed.
- **erodible**
  a term used to describe a soil that is vulnerable to erosion.

- **erose**

  1. ragged, with irregular random indentations.
  2. having an irregularly jagged margin, as if gnawed, applied to a leaf or other plant part.
  3. (L. *erodere*, to wear away) having small irregular notches in the margin, as if gnawed.

- **erosion**

  the wearing away of the land through the action of moving water, wind or other geological agents.

- **erosive**

  used to describe water that has enough force to cause erosion.

- **erucic acid**

  the major fatty acid found in rapeseed oil used for industrial purposes. Varieties of oil with low erucic acid content such as canola oil have nutritional value.

- **escape**

  an exotic species spreading from cultivation and becoming established in natural habitats. A plant that has escaped from cultivation and *naturalised* more or less permanently (i.e. Queen Anne's lace is an escaped carrot).

- **espalier**

  1. a trelliswork of various forms on which the branches of fruit trees or fruit bushes are extended horizontally, in fan shape etc., in a single plane. The object it to provide better air circulation and sun exposure for the plants.
  2. a tree or plant trained on such a trellis or system. Trees trained to espaliers aren't as liable to suffer abrupt variations of temperature as are wall trees.
  3. to train on or protect by an espalier, as a tree or trees.

- **espathate**

  lacking a spathe.

- **esquamulose**

  lacking minute scales, not *squamulose*.

- **essential flower parts**

  the stamen and pistil organs of the flower that are required for pollination.

- **ESTS (expressed sequence tags)**

  DNA sequence information developed by partial sequencing of a large number of randomly selected cDNA clones. These sequence databases describe the kinds of genes and expression levels in a particular tissue.

- **estuarine**

  associated with estuaries or river mouths, usually in brackish conditions.

- **estuary**
(L. *aestuarium*, part of the sea-coast over which the tide ebbs and flows, from aestus, the tide) an inlet or arm of the sea, especially the wide mouth of a river, where the tide meets the current. pl. estuaries.

- **etagere**
a series of open shelves for growing and displaying plants, a tiered plant stand.

- **etectate**
describing pollen grains interpreted as having lost their tectum during evolutionary development. See **atectate, intectate, tectum.**

- **ethion**
a synthetic insecticide effective against sucking and chewing insects, especially whiteflies and mites.

- **ethnobotany**
the study of the uses of plants by various human cultures.

- **etiolate**
1. leaves and/or stems that are unnaturally long, often with a change of colour from the natural shade of green to yellow or white, due to insufficient light.
2. to blanch or whiten a plant or part of a plant by excluding sunlight. This is often done with endive, celery and cauliflower.

- **etiolation**
1. the process of causing a plant to whiten by excluding sunlight.
2. pale, sickly, excessive growth caused by insufficient light, incorrect feeding or overcrowding.

- **eu-**
a prefix for true.

- **eucyclic**
having a regular alternation of parts, applied to flowers in which the petals, stamens etc. are equal in number in each whorl and alternate with one another.

- **euintine**
synonym of endintine.

- **eukaryote**
organism the cells of which have a defined nucleus and internal membranes.

- **eukaryotic**
having membrane-bound organelles.

- **euphorbus**
physician to King Juba II (ca. 50 BC- 19 AD) of Mauretania. King Juba was said, by Pliny the Elder (23-79 AD), to have named a plant (probably *Euphorbia resinifera*) that he discovered on Mount Atlas in Morocco as 'Euphorbea' after his physician.

This is generally considered to be the origin of the name of the genus *Euphorbia*. A variety *E. regis-jubae* of *Euphorbia obtusifolia* commemorates King Juba himself.

- **euphotic zone**
  that upper layer of the ocean that receives sufficient light for photosynthesis to occur, to depths of perhaps 180 feet in Monterey Bay.

- **euphyllophytina**
  clade comprising the seed plants, sphenopsids, ferns and *Ps ilophyton*, synapomorphies for the clade include among others a basically helical arrangement of small, pinnule-like vegetative branches, recurvation of branch apices and paired sporangia grouped into terminal branches.

- **euryhaline**
  tolerant of range of salinities.

- **eurypalynous**
  describing plant taxa characterised by possession of a great diversity of palynomorphs.

- **eusporangiate**
  having the sporangium develop from a great amount of leaf tissue as opposed to only one or a few cells.

- **eusporangium, eusporangiate**
  eusporangium, sporangium that arise from a group of superficial cells, this is the primitive character state in vascular plants, compare to leptosporangium.

- **eustele**
  1. a stele type, in which the vascular bundles are organised into strands of xylem flanked by strands of phloem on the outside, in transverse section the eustele appears as a ring of discrete vascular bundles.
  2. a dissected siphonostele with phloem only to the outside of the xylem.
  3. a ring of vascular bundles surround a pith.

- **eutectum / eutectate**
  describing a pollen grain with a continuous tectum.

- **eutrophic**
  nutrient-rich. (Gr. *ew*, well *trophe*, nourishment) the gradual increase in nutrients in a body of water. Natural eutrophication is a gradual process, but human activities may greatly accelerate the process.

- **eutrophication**
  the process of becoming eutrophic.

- **evaporation**
  the process of conversion of a liquid into a vapour or gaseous form. This change of state absorbs energy- the latent heat of vaporisation.

- **evapotranspiration**
  the loss of water from the soil by evaporation and plant transpiration. Or water on the surface taken away by water, wind, air and elements.

- **even-pinnate**
  said of compound leaves having an even number of leaflets, this is usually easily determined because there is a pair terminally.

- **everbloomer**
  a plant that blooms continuously throughout the growing season.

- **evergreen**
  1. remaining green and leafy throughout the winter or resting season.
  2. opposite of deciduous, i.e. evergreen plants do not shed all their leaves annually, but shed and grow new leaves continually; most conifers are evergreen, but the evergreeen habit is not restricted to conifers.
  3. keeping its leaves all year long, although losing some of the older leaves regularly throughout the year. Semi-evergreen plants only lose some leaves or lose older leaves only when new ones develop.
  4. bearing green leaves through the winter.

- **everlastings**
  this describes plants whose flowers maintain their colour, brightness and form long after being gathered.

- **evolution**
  a process of selection originally described by Charles Darwin in 'The Origin of Species', whereby organisms possessing one or more characteristics conferring increased fitness, tend to survive and proliferate more successfully than similar organisms without these features. Natural selection therefore occurs for organisms possessing these desirable characteristics and their numbers increase disproportionately within the population. Evolution generally proceeds very slowly, by selection for very small differences in characteristics. However, occasional introduction of large beneficial (in terms of fitness) changes into the gene pool leads to *punctuated evolution* where major changes occur over relatively short times compared with the geological record. The *Cactaceae* appear to be a group of plants which have diversified relatively recently, with few, if any, authentic fossils. Compare with creation.

- **evolutionary grade**
  same as paraphyletic groups-howing similarities in morphology, ecology or life history.

- **evolutionary systematics**
  same as synthetic systematics. A way to determine natural relationships of organisms by studying a group in detail and comparing degree of similarity; evolutionary systematics does not have an explicit methodology, but rather relys on the expertise of authorities very familiar with the group in question, compare to phenetics or numerical taxonomy and phylogenetic systematics or cladistics.

- **exalate**
  lacking wings or wing-like appendages. Not alate.

- **exalbuminous**
  without endosperm, applied to seeds. Also endospermous, exendospermic.

- **exanthem**
  a blotch on the surface of a leaf etc. A copper-deficiency disease of plants, prevalent in citrus and olive and characterised by die-back and blotches on leaves and fruit.

- **exarch**
  a type of xylem maturation in which protoxylem is external to metaxylem and development prodeeds centripetally (from the outside in), for comparison see **endarch and mesarch.**

- **exchangeable**
  a term used to describe ions in the absorbing complex of soil that can be exchanged with other ions.

- **exchangeable sodium**
  sodium that is attached to the surface of soil particles and that can be exchanged with other positively charged ions in the soil solution.

- **excrescence**
  1. an outgrowth or protrusion from the surface.
  2. (L. *excrescere*, to grow out. *ex*, out, *crescere*, to grow) a normal outgrowth. A disfiguring addition.

- **excurrent**
  (L. *excurrere*, to run out, project) projecting beyond the tip, as the midrib of a leaf or bract.

- **exfoliate**
  to peel off in layers or flakes, as the bark of certain trees. vt. (L. *ex*, out; *folium*, leaf) peeling off in thin layers, shreds or plates, as the bark of some trees.

- **exine**

  1. the outermost layer of the wall of pollen and spores; made of sporopollenin; compare to intine.
  2. the outer layer of the wall of a spore or pollen grain.
  3. (L. *ex*, out of, out) the outer of two layers forming the wall of certain spores (as pollen grains) - called also exosporium.
  4. the outer layer of the wall of a palynomorph, which is highly resistant to strong acids and bases and is composed primarily of sporopollenin.

- **exintine**

  an outer (pectic) layer of the intine, lying below the nexine and which stains positively with alcian blue.

- **exitus**

  the site of exit of the pollen tube from the sporoderm. See **aperture.**

- **exocarp**

  1. the outer layer of the covering of a fruit when it consists of two dissimilar layers.
  2. (Gr. *exo*, without *karpos*, fruit) the outer layer of the wall of a matured ovary.

- **exoexine**

  synonym of sexine.

- **exogamous**

  pollination between genetically different plants.

- **exogen**

  a plant in which the growth of the stem is in successive concentric layers, such as most trees.

- **exogenous**

  growing by additions on the outside, such as the rings of a tree trunk.

- **exogerminal**

  describing an aperture formed in the outer wall layer of the exine.

- **exospore**

  the outer layer of a spore wall.

- **exospore or exine**

  outer spore wall layer.

- **exosporic**

  gametophyte development outside the spore wall, i.e. the development of a free-living, multicellular gametophyte.

- **exosporium**

  synonym of exine.

- **exotic**

  introduced from a foreign country or a different region. An exotic plant is one not native to the place where it is growing such as Japanese Honeysuckle, which has naturalised in the northeastern states or Eucalyptus trees, which have been naturalised on the West Coast.

- **expressed**
  genes for which there is current expression of the mRNA ('turned on'). For example, a gene may be expressed in a particular tissue or at a particular developmental stage but not at other times or in other tissues.

- **exsert**
  1. vt. (L. *exserere*, to strech out) to put forth, to thrust out, to protrude.
  2. of stamens or pistils exceeding the length of the tepals.
  3. sticking out, extending beyond (some enclosing part).
  4. projecting beyond, sticking out.

- **exstipulate**
  (L. *ex*, private *stipula*, a stalk, stem) having no stem.

- **exsuccous**
  dry, without juice, sap or moisture, withered.

- **extant**
  currently existing. Living now.

- **extinct**
  of a species which has completely died out. species can also become 'extinct in the wild' but persist in botanical or other collections. No longer existing, not living anymore.

- **extinction**
  the process leading to complete loss of a species.

- **extrafloral**
  of a plant part that doesn't form part of a flower.

- **extrorse**
  facing away from the direction of the growth (in reference to the anther of a flower).

- **eye**
  the centre of a flower when it is a different color from the petals. An undeveloped bud on a tuber, as on a potato. A cutting with a single bud.

- **f1**
  of progeny derived from a first generation cross between distinct individuals.

- **f2**
  of progeny from self fertile F1 hybrids.

- **facultative**
  not restricted to a particular way of life. Exhibiting some ability or function under some environmental conditions but not under others.

- **falcate**
  1. strongly curved or sickle-like. Hooked, curved like a scythe or sickle, falciform.

2. (L. *falx*, a sickle) curved like a sickle.
3. sickle-shaped, curved with parallel edges.

- **fallow**

plowed and left unseeded, left for a considerable time unworked or unseeded after tillage. untilled, uncultivated.

- **falls**

the drooping or horizontal petals of irises.

- **false indusium**

a folded leaf margin protecting the sorus.

- **false veins**

small vein-like areas of thick-walled cells in the leaves of some lower vascular plants.

- **family**

1. a taxonomic grouping of similar genera.
2. this consists of genera (singular is *genus*, go see) that resemble one another in certain broad characteristics. The family usually consists of numerous genera, though some only consist of one. The family name usually ends with -aceae or -ae. For example, Compositae, which is the largest family of plants, consists of 950 genera and 20,000 species. Asters, Sunflowers, Mums and Dahlias are all well-known members of that family and these genera contain numerous species.
3. a group of related genera taxonomically classified within an order.

- **fan-training**

a method of training a tree or vine on a trellis or wall in a way that causes the branches radiate from the trunk at regular intervals and at continually smaller angles, the lower branch on each side being horizontal.

- **farina**

a whitish mealy or flour-like covering. Hence adj. *farinose, farinaceus.*

- **farinaceous**

(L. *farina*, meal) containing flour, starchy, mealy.

- **farinose**

1. covered with mealy powder (i.e. leaves of certain plants.
2. full of meal; mealy; covered with a white, powdery substance.

- **fasciarate (adj)**

band-shaped. Long and narrow with untapering margins, with a length, breadth ratio of 12+:1.

- **fasciate**

banded or compacted together. Exhibiting fasciation. Also fasciated.

- **fasciation**
  1. malformation of a stem or other part usually with lateral flattening, cresting.
  2. a malformation in plants, in which a stem or branch becomes expanded into a flat, ribbon-like shape, as if several stems were joined together lengthwise.
- **fascicle**
  1. a close cluster or bundle, as of leaves, stems or flowers. Sometimes limited in use to a compacted cyme.
  2. (L. *fasciculus*, small bundle) a small bundle or tuft, as of fibers, leaves etc.
  3. a close cluster of flowers.
- **fascicular cambium**
  arises within vascular bundles.
- **fastigiate**
  with erect branches and close together.
- **fastigiate trees**
  trees grown for their narrow, columnar shape.
- **fastigium (pl. fastigia, adj. fastigiate)**
  cavity in a colporate grain, appearing as a separation of the inner part of the exine from the domed sexine in the region of the endoaperture. Example: Hypericum (hypericaceae).
- **faveolate**
  honeycombed, alveolate. In botany, also favose.
- **feed**
  food for livestock, usually consisting of grains such as corn, rye and barley.
- **felt**
  a very thick layer of hairs or fibres.
- **female**
  a plant that produces fruit. The plant that bears the pistil and receives the pollen or fertilising element of the male plant or the analagous organ in cryptogams. Pertaining to the kind of plants that produce fruit; pistil bearing, pistillate, producing pistillate flowers.
- **fen**
  (ME. *fen, fenne*; AS. *fen, fenn*, a marsh, bog, fen) low land covered wholly or partially with water but producing sedge, coarse grasses or other aquatic plants; boggy land. A moor or marsh, a plant community on alkaline, neutral or slightly acid peat.
- **fenestrate (adj.)**
  1. describing a class of pollen grains characterised by large, window-like spaces lacking a tectum.

2. (L. *fenstra*, window) a type of leaf anatomy with small perforation or transparent spots. Confined to a few tropical monocotyledons which grow on the island of Madagascar.

- **fenestration (n)**
  the pattern of translucent spots that occurs on and around the lids of certain pitchers of the genera Cephalotus and Sarracenia.

- **fernery**
  a place where ferns are growing, often in a greenhouse, a plantation of ferns.

- **ferric iron**
  an oxidised or high-valency form of iron.

- **ferrous iron**
  a reduced or low-volence form or iron.

- **ferruginous**
  rust-coloured or reddish-brown.

- **fertile**
  1. a plant with functional reproductive organs, producing pollen capable of fertilisation and seeds or spores capable of germination.
  2. producing abundantly, capable of sustaining abundant growth, as fertile soil.
  3. capable of producing fruit, having a functional pistil, bearing functional pollen etc.
  4. capable of fertilising, as an anther with well-developed pollen.
  5. capable of engaging in the process of fertilisation.

- **fertilisation (n)**
  the process of creating a zygote by fusing two gametes, usually accomplished by pollination.

- **fertiliser**
  a mix of essential nutrients supplied to plants to improve their growth or to correct a specific deficiency. For commercial fertilisers the proportions of Nitrogen:Phosphorous:Potassium are marked on the package as N:P:K ratios.

- **fertility**
  the quality of soil that enables it to provide compounds in adequate amounts and in proper balance for the growth of specified plants.

- **fertilisation**
  the act of rendering land fruitful or productive, the application of fertiliser. The process by which pollen reaches the ovules and assures the production of fruit, also, the analogous process in cryptogams.

- **fertiliser**
  a natural or man-made material added to soil to supply nutrients to plants.

- **fertiliser grade**
  an expression that indicates the percentage of plant nutrients in a fertiliser.

- **fertiliser injector**
  a device connected to the faucet end of a garden hose. It contains about a pint of liquid or water soluble fertiliser and mixes it at a predetermined rate with water flowing through the hose.

- **fibre**
  1. a long-walled plant cell which is often dead at maturity. Fibres impart elasticity, flexibility and tensile strength to plant structure
  2. long, straight and thin, often occurring in bundles. Sometimes called 'extraxylary fibres' to distinguish them from xylary fibres, which look similar, but have a different evolutionary origin.

- **fibre tracheid**
  a cell in the xylem that is intermediate between a tracheid and a libriform fibre.

- **fibril**
  a small fibre or a root hair.

- **fibrillose**
  furnished with or made up of fibrils.

- **fibrous root**
  finely divided root system without obvious thickening or an enlarged central root. A fine, densley branching root, which absorbs moisture and nutrients from the soil.

- **-fid (suf)**
  pointed- lobed, but only narrowly joined in relation to the size of the segments - eg: 2-fid, 3-fid.

- **field capacity**
  the highest amount of moisture remaining in a soil after free water has been allowed to drain away. Expressed as a percentage of oven-dry weight of soil or other convenient unit.

- **field moisture**
  the water that soil contains under field conditions.

- **filament**
  1. stalk of a supporting anther.
  2. the stalk of a stamen that bears the anther.
  3. stalk-like portion of the stamen.
  4. a thread-like structure

- **filiciform**
  in the shape of a fern or fern frond.

- **filiform**
(L. *filum*, thread; *forma*, shape) thread-like, long and very slender. In the form of a filament, coarser than capilliform.

- **film water**
water held on the surface of the soil particles that doesn't drain away.

- **fimble**
the male hemp plant or the fiber made from it. Also called fimble hemp.

- **fimbria**
a fringe-like border; the peristome of a moss.

- **fimbria (pl. fimbriae, adj. fimbriate)**
long, hair-like appendages.

- **fimbriate**
1. fringed, bordered with hairs or with slender processes, as a fimbriate petal. Also fimbriated.
2. (L. *fimbriatus*, fringed) cut into regular segments and appearing fringed at the margins.
3. fringed with hairs.

- **fine sand**
soil particles measuring from 10 to .25 millimeter.

- **fine-textured soil**
a clayey soil containing 35% or more clay.

- **firming**
a term used to indicate light-packing of soil around roots.

- **first bottom**
the normal flood plain of a stream.

- **fischer's law/rule**
the name given to the wide-spread arrangement in developmental tetrads of tri-aperturate pollen grains whereby the apertures form in pairs at six points in the tetrad. Example: Ericaceae. See **Garside's law/rule.**

- **fish meal**
an organic fertiliser made of fish.

- **fissura (pl. fissurae)**
a sharp, straight split that appears during germination in certain inaperturate pollen grains. Examples: Taxodium (Taxodiaceae), Cupressus (Cupressaceae).

- **fistula**
(L. *fistula*, pipe) pathological or artifical pipe-like opening. Water-conducting vessel - alt. trachea.

- **fistulose**
same as fistulous.

- **fistulous**
having the form or nature of a fistula. Cylindrical or terete,

hollow and fully enclosed, i.e. capped at both ends.

- **fixation**
the conversion of a soluble material form to a relatively insoluble form.

- **flabellate**
fan-shaped. Also flabelliform.

- **flabelliform (adj)**
fan-shaped.

- **flaccid**
1. limp, flabby or drooping.
2. lacking strength or weak.
3. (L. *flaccidus*, flabby) weak, limp, soft or flabby. Leaves that do not have enough water and are about to wilt or are wilting.

- **flagelliform (adj)**
whip-shaped, most often referring to roots or stems.

- **flagellum**
1. (Pl. flagella)- a whip-like hair, usually motile in lower plants and plankton. Hence adj. *flagelliform* = whip-like, *flagellate* = a single cell form of plankton with a flagellum.
2. a runner or stolon of a plant that roots and forms new plants at the nodes (i.e. the strawberry).
3. a young shoot or twig.

- **flagellum/a**
whip-like extension of the protoplasm for propelling the cell.

- **flange**
a general term, used to describe equatorial extensions of spores. See **capsula, cingulum, corona, patina, zona.**

- **flash tape**
a breeze-activated, metalised plastic tape that produces bursts of light in response to breezes, suspended over crops to scare away birds.

- **flat**
a container for holding packs of plant starter cells. Also called seed flat.

- **flavescens**
becoming yellow.

- **flavispina**
with yellow spines.

- **flexuous**
wavy or zigzag.

- **floating row cover**
this is a lightweight fiber sheet, water and air penetrable, that is placed over a row of plants for protection from heat, cold or insects.

- **floccose**
1. (of plants) having tufts of short soft woolly hairs.
2. covered with tufts of soft, woolly hairs.

3. (L. *floccus*, a lock of wool) said of pubescence which gives the impression of irregular tufts of cotton or wool.
4. having tufts of fine hair, woolly.

- **flocculate**
  to clump together individual tiny soil particles.

- **flocculose**
  finely floccose. Also flocculous.

- **floccus**
  a small tuft of woolly hairs.

- **flood irrigation**
  a shallow flood of water over level soil.

- **flora**
  1. all the plant species in an area. A book or treatise on plant species from a particular region.
  2. plant life in general, often in contrast to fauna. The plants native to or adapted to a particular region, habitat etc. or belonging to a particular period.

- **floral**
  containing, belonging to or relating to a flower, pertaining to flowers in general or made of flowers.

- **floral envelope**
  the *perianth* of a flower.

- **floral organs**
  modified leaves specialised for reproduction.

- **florescence**
  1. blooming or the unfolding of a flower, the state of being in bloom.
  2. (L. *florescence*, to begin to flower) bursting into bloom, alt. anthesis.

- **florescent**
  being at the point or stage of florescence.

- **floret**
  1. a small flower, especially as one component of a dense multiple flower cluster.
  2. a single flower in a head of many flowers.
  3. (L. *flos*, flower) one of the small individual flowers of a crowded inflorescence such as capitulum; flower with lemma and palea, of grasses, alt. floscule.
  4. each of the small flowers making up a compound flower.

- **floricane**
the stem at flowering and fruiting stage (of a bramble, Rubus).

- **floriculture**
the cultivation of flowers or of flowering plants, usually commercially.

- **floriculturist**
a person who specialises in growing flowers.

- **floridean starch**
an algal reserve resembling glycogen or amylopectin.

- **floridean starch**
branched glucan polymer, storage produce of Rhodophyta present as granules outside the chloroplasts.

- **floriferous**
flower-bearing, usually used to indicate profuse flowering.

- **floriform**
in the form of a flower.

- **floriparous**
this pertains to plants that normally bear fruit but instead produce secondary flowers.

- **florist**
one who sells flowers and ornamentals.

- **florist's flower vase**
a tapered metal bucket, slightly flared at the top, is used to hold cut flowers.

- **floscular**
same as *discoid*, as applied to disk flowers in the family Compositae. Composed of florets. Also flosculous, flosculose.

- **floscule**
(L. *flosculus*, little flower) a small flower or a floret.

- **flower**
1. the complex reproductive organ of an angiosperm plant which may include sepals, petals, stamens and pistil. Flowers may include either male or female parts or both. The colourful parts of a flower and its perfume attract pollinators and guide them to the nectary, usually at the base of the flower tube. Pollination usually occurs incidentally to foraging for nectar or pollen as food. Xerophytic plants may have flowers that are produced and open in response to rain and produce seed within a few weeks of pollination, thus completing their life cycle. Species with tubular red flowers have often evolved to attract humming birds and other avian pollinators. Some night-flowering species have specialised to attract bats with foetid fleshy blooms or moths with large strongly-perfumed flowers.

2. condensed reproductive shoot of flowering plants. The diversity of flower form comes from variation of the general four-whorled structure.
3. the part of the plant that ordinarily contains the reproductive organs, which are usually surrounded by colourful *petals* and *sepals*. A flower containing only *stamens*, is male or staminate. A flower containing only *pistils*, is female or pistillate. When a flower contains both pistils and stamens it is called a bisexual or perfect, flower. If it also has a *corolla* and *calyx*, it is called a complete flower.
4. flower is also generally used to describe any blossom or inflorescence or any plant of which the blossom is the essential feature. A plant grown for its floral beauty.
5. (ME, *flowre*, *flour*, *flur*; OFr., *flor*, *flur*, *flour*, from L. *flos*, *floris*, a flower) the part of a plant containing or consisting of the organs of reproduction, either together in a monoclinous flower or separate in male and female flowers.

- **flower arrangement**

a combination of individual cut flowers to create an attractive decoration.

- **flower head**

a dense arrangement of flowers arising from a common point as in e.g. *Umbelliferae* or *Compositae*.

- **flower heads**

a mass of small flowers or florets that together appear to form a single flower (e.g. Daisy), a form of inflorescence giving the effect of a single flower, consisting of a compact cluster of florets *sessile* upon the shortened summit of the axis.

- **flower press**

a contraption in which flowers may be dried and preserved. A flower press consists of layers of cardboard and blotting paper sandwiched between two wooden boards, which are held together in the corners by adjustable screws.

- **flower ring**

a wire loop supported by stakes to prop and contain a cluster of weak-stemmed flowers such as bachelor buttons or campanulas.

- **flower stalk**

a peduncle or pedicel. The ordinarily leafless part of a stem or branch that bears a flower cluster or a single flower.

- **flower support**
  a stake, usually made of plastic coated wire, to prop or brace a single weak-stemmed plant such as lily or gladiolus. Also called loop stake.

- **floweret**
  a floret, sometimes refers to one of the segments of a cauliflower head.

- **flowering plant**
  an angiosperm.

- **fluorescent tube**
  a type of light bulb that emits ultraviolet light. Fluorescent lamps are used to start and maintain plants indoors.

- **fluorescent-light gardening**
  the process of growing plants under artificial lights.

- **foetid**
  with an offensive stinking odour.

- **foliaceous**
  1. being or resembling a leaf. In botany, having the texture or form of a leaf, as a foliaceous sepal or bract.
  2. (L. *folium*, leaf) thin and leaf-like, bearing leaves.
  3. resembling a leaf in texture or form.

- **foliage**
  the leaves of plants. The mass of growing plant leaves in their natural form and condition or a cluster of the same. Leafage.

- **foliage plant**
  a plant grown for its attractive foliage rather than for its flowers, as the various kinds of Coleus and Philodendron.

- **foliar**
  consisting of or pertaining to leaves.

- **foliar feeding**
  feeding plants by spraying plant food on their leaves.

- **foliate**
  leafy, covered with leaves or leaflets (i.e. foliate stalk).

- **foliation**
  the leafing of plants. The disposition of the beginning leaves within the bud.

- **foliicolous**
  growing upon leaves or parasitic on leaves, as many fungi.

- **foliolate**
  pertaining to, consisting of or having leaflets. Used in combination, as bifoliolate, having two leaflets.

- **foliole**
  a leaflet, a small part resembling a leaf.

- **foliose**
  leafy or resembling a leaf. Also folious.

- **follicle**
  1. a dry dehiscent, many-seeded fruit derived from one carpel and splitting along one suture.
  2. a dry, one-celled seed vessel consisting of a single carpel and *dehiscent* only by the *ventral* suture, as in the milkweed and larkspur.
  3. (L. *folliculus*, small sac) a dry dehiscent fruit formed of one carpel and dehiscing along one side.

- **foot layer**
  the inner layer of the ectexine.

- **footstalk**
  the stalk or petiole, of a leaf or the peduncle of a flower.

- **foramen (pl. foramina, adj. forate, which is an abbreviation of foraminate)**
  synonym of pore.

- **foraminate**
  contains several round perforations.

- **forcing**
  the practice of bringing a plant into growth or flower (usually by artificial heat or controlling daylight) at a season earlier than its natural one.

- **forcing house**
  a hothouse used for forcing plants.

- **forcing pit**
  a pit of wood or masonry, sunk in the earth, for containing fermenting materials to produce bottom heat in forcing plants.

- **forest**
  a plant community dominated by trees or other large plants e.g. large Aloes.

- **form**
  1. the smallest degree of consistent differentiation between plants which may be formally named e.g. spine length or flower colour.

2. a botanical category ranking below a variety and differing only trivially from other related forms, such as in flower colour.

- **form taxon**
  binomial name (genus and species) given to a fossilised plant organ when it is found in isolation, i.e. when the taxonomic affinities of the organ are not known with certainty. For example spore and pollen taxa have their own binomial names, since it is rarley known which fossil genus may have produced them.

- **fornicate**
  overarched fornices, as the throat of the corolla of the forget-me-not. Also forniciform.

- **fornix**
  (plural- fornices) A small, arching scale or appendage in the throat of the corolla of some plants, such as in forget-me-nots.

- **fossaperturate (adj.)**
  refers to an equatorially aperturate, lobate pollen grain with the apertures in the indentations between the lobes.

- **fossil**
  any evidence of past life. Any remains of any once living organism preserved in the Earth's rocks

- **fossula**
  synonym of areola. A feature of ornamentation consisting of an elongated, irregular groove in the surface. See **foveola.**

- **fountain growth**
  multiaxial growth usually of many filaments in which cell divisions occur.

- **fovea**
  pit or depression containing the sporangium in the leaf base of *Isoetes.*

- **foveola (pl. foveolae, adj. foveolate)**
  a feature of ornamentation consisting of more or less rounded depressions or lumina more than 1μm in diameter. The distance between foveolae is greater than their breadth. See **lumen, punctum.**

- **fragilis**
  fragile, easily broken.

- **fragipans**
  the hard and brittle layers in soil that roots and water can't penetrate because it is so dense and compact.

- **fragmentation**
  production of new individuals from fragments of original individual.

- **frame**

a garden frame is a bottomless box with a transparent top that is used to provide favorable conditions for the plant's growth and to cover them during harsh weather. The top is made of plastic or glass set in a wooden or metal frame. To allow for proper ventilation, the top is formed with one or two moveable sections called sash. The sashes are moved by sliding them up and down or, if they are hinged, by raising their fronts. Their are unheated and heated frames and they are both good for raising seedlings ahead of the outdoor season and for growing early crops of Lettuce, Cauliflower, Radishes, Corn, Snapdragons and other vegetables and flowers.

- **free**

1. used to describe the condition of a substance within a mixture when it isn't chemically combined with other components of the mixture.
2. no veins uniting to form a network.

- **free nuclear division**

mitotic division of nuclei without accompanying cytokinesis, i.e. nuclei divide in a common cytoplasm, the cells walls only forming around each later.

- **fresh weight**

same as wet weight. Generally not a useful measurement for aquatic plants.

- **friable**

rich, loamy soil that is easily worked and loose. Soil of a crumbly texture, able to be worked easily. Friable soil crumbles easily when worked with a spade, fork, rake etc. The opposite is a tenacious, clayey soil. The particles of this soil cling together forming clods that are difficult to pulverise. Nonfriable soil can be enhanced by the following methods- Mixing in organic matter such as manure, leaf mold, peat moss and compost; turning under cover crops or green manures such as Winter Rye, Ryegrass and Buckwheat and adding sand or gritty coal cinders, liming and the use of synthetic soil conditioners.

- **frond**

leaf of a fern.

- **frond**

a leaf of a palm or fern.

- **frondage**

a group of fronds.

- **frondent**

leafy.

- **frondescence**
  the period or state of coming into leaf. Leafage or foliage.

- **frondescent**
  coming into leaf.

- **frondiferous**
  producing fronds or leaves.

- **frondlet**
  a small frond.

- **frondose**
  bearing or resembling fronds.

- **frost**
  1.that state or temperature of the air that occasions freezing of water.
  2. the state or condition of being frozen of the surface of the ground when frost extends to a depth of 10 inches.

- **frost control**
  the act of minimising frost damage to plants by covering them, watering them lightly or placing them in a protected area.

- **frost-free days**
  the number of days in a growing season in any given area.

- **fructescence**
  the time fruit matures.

- **fructiferous**
  bearing or producing fruit.

- **fructification**
  the production of fruit by a plant or fruiting.The ripened ovary and its appendages. The fruit of a plant.

- **fructify**
  to bear or produce fruit or to make fruitful.

- **fruit**
  1. the mature ripened reproductive body (ovary) of a seed plant which encloses the seed in fertile species of plant. True fruits develop from the ovary wall. Many fungi, myxobacteria (slime moulds) and lichens produce a fruiting body, a macroscopic reproductive structure enclosing the spores. Fungal fruiting bodies often have distinctive sizes, shapes and colours characteristic of each species.

  2. any product of vegetable growth useful to humans or animals (i.e. grapes, figs, corn, cotton, flax and all cultivated plants). There are many kinds of fruit and they differ greatly in character and degree of complexity.

3. the reproductive product of a plant; the seed of plants or the part that contains the seeds.
4. an edible, juicy product of a plant, usually covering the seeds (i.e. apple orange, a berry, a melon etc.)
5. the matured ovary of a plant, consisting of the seeds and their pericarp (coating) and including whatever may be incorporated with it. Also, the spores of cryptogams and the organs accessory to them. (verb).
6. to produce fruit or to come into bearing.
7. a mature ripened ovary. In practice, most plant parts that contains seeds are fruits.
8. (Fr. *fruit*, from L. *fructus*, fruit) the developed ovary of the flower containing ripe seeds, whether fleshy or dry, often used to include other associated parts such as a fleshy receptacle, then called a false fruit.

- **fruit dot**

the *sorus* of ferns.

- **fruit tree**

a tree grown for its fruit or a tree whose main value consists in the fruit it bears (i.e. cherry tree, apple tree, pear tree etc.).

- **fruit trees, dwarf**

describes genetic dwarf trees and trees grafted onto dwarf roots. For example, peaches that are genetic dwarfs produce full size fruit on trees that hardly ever grow more than 3 feet high. Fruit trees grafted onto dwarf rootstock, such as citrus or apples, produce normal fruit on trees that only grow up to 8 feet high. Many grafted fruit trees are also sold as semi-dwarf, growing to about 10 feet high. Dwarf fruit trees are great for growing in suburban yards, containers and in areas with little space.

- **fruit, compound**

a fruit that is made of several ovaries.

- **fruiting body**

a specialised organ for producing spores, as in mosses and mushrooms.

- **fruitlet**

a small fruit or unit of a collective fruit.

- **frumentaceous**

having the character of or resembling wheat or other cereal.

- **frustillum (pl. frustilla, adj. frustillate)**

synonym of areola.

- **frustule**

siliceous cell wall of a diatom.

- **frutescens**
  shrubby. Having the appearance or habit of a shrub, shrubby, as a frutescent stem.
- **fruticose**
  resembling a shrub or shrubby, as a fruticose stem.
- **fruticulose**
  resembling a small shrub.
- **fucoidan**
  water-soluble, sulfated polysaccharide in Phaeop' 7cean cell wall.
- **fucosan vesicle**
  small, refractive particle or physode in cell of Phaeophycea, containing tannin-like compounds.
- **fugacious**
  withering away early, ephemeral.
- **fulgid**
  shining.
- **fulgida**
  glittering, shining.
- **fulvous**
  yellow or tawny or yellow flecked with a mixture of grey or brown.
- **funaliform (adj)**
  in the form of rope, composed of rough fibres.
- **functional food**
  a functional food is 'consumed as part of a usual diet that is similar in appearance to or may be, a conventional food and is demonstrated to have physiological benefits and/or reduce the risk of chronic disease beyond basic nutritional functions'.
- **functional genomics**
  an area of study that determines what trait or function a gene encodes.
- **fungal**
  pertaining to, characteristic of or consisting of a fungus or fungi. Also fungous.
- **fungal spores**
  unicellular or multicellular, reproductive or distributional cells developing into a number of different phases of the complex life cycles of the fungi.
- **fungi/ fungus**
  a division of diverse plants that lack chlorophyll and do not photosynthesise. Included are mushrooms, molds, rusts, smuts, and yeasts. Some are edible,

such as Boletus. Many are plant diseases, such as powdery mildew.

- **fungicide**
an agent that destroys fungi. A material used to control fungal disease on plants, plant parts and soil.

- **fungiform (adj)**
mushroom-shaped.

- **funicle**
the stalk of an ovule or seed. The thread like tissue connecting the developing seed, via the hilum, to the fruit. (Latin: *funiculus* = little rope)

- **funiculus**
1. the stalk of an ovule. The stem of a seed or its case. Also funicle.
2. the stalk that attaches a seed to the inside of a fruit.
3. (NL. from L. *funis*, a small rope) the stalk of an ovule.

- **funnel form**
having the shape of a funnel, Applied to the corolla of a flower in which the tube enlarges gradually from below, as the morning glory; infundibuliform.

- **furcate**
forked.

- **furcellaran**
sulfated polysaccharide gel (like carrageenan) found in cell wall of red alga Furcellaria.

- **furfuraceous**
coated with bran-like, flaky particles, scurfy.

- **furrow**
a trench in the earth, especially one make by a plow. To cut a furrow in, make furrows in, plow. A common word for an elongate aperture.Examples: colpus, sulcus, tenuitas.

- **furrow membrane**
synonym of colpus membrane.

- **furrow weed**
a weed growing on plowed land.

- **furrowed**
with parallel longitudinal grooves or channels. Having longitudinal ridges or grooves, as a furrowed stem.

- **fusain**
fossil charcoal a component of coal and sedimentary rocks characterised by black colour, silky luster and fibrous texture. Carbonaceous material derived from decaying vegetation or wood.

- **fused**
joined and growing together.

- **fusiform**
1. tapering towards each end, football-shaped.
2. like a spindle, as the radish. Also fusate.
3. (L. *fusus*, a spindle; *forma*,) thick.

- **fusiform initial**
  an elongated cell in the vascular cambium, produces elements of the axial system.

- **fusion cell**
  cell produced by fertilisation.

- **galbulus**
  a spherical cone formed of thick scales with a narrow base, as in the Cypress or berry-like with fleshy scales, as in the Juniper.

- **galea**
  a helmet-shaped part of a calyx or corolla, as the upper lip of a ringent or *labiate* corolla.The 'helmet' formed by the dorsal sepal or fusion of dorsal sepal and petals in some Orchids.

- **galea (pl. galeae, adj. galeate)**
  a relatively large element of the outer wall of a spore consisting of a sharply tapering spine and a broad bulbous base.

- **galeate**
  having a galea, helmet-shaped (i.e. the helmet-shaped upper sepal of Monkshood).

- **galeiform**
  helmet-shaped.

- **gall**
  an abnormal plant growth on leaves, branches or roots that is caused by irritation from an insect, a fungus or a virus.

- **galvanotropism**
  bending, growth or movement of a part under the influence of an electric current.

- **gametangium**
  cell in which gametes form.

- **gamete**
  1. the male and female sexual cells that combine at fertilisation to form the zygote.
  2. a mature sex cell (i.e. a sperm or an egg) possessing a haploid chromosome set and capable of developing into a new individual by fusion with another gamete.
  3. sexual cell capable of uniting in fertilisation with compatible cell to form zygote.
  4. a specialised cell which unites with another to form a zygote.

- **gametophyte**
  1. the haploid phase (n=1) of a life cycle on which gametes are produced.
  2. the generation or stage of a plant producing gametes, existing only as a rudimentary structure in seed plants and as distinguished from the sporophyte or asexual form.

Fern Gametophyte with Attached Young Sporophyte (Two views of the same plant)

3. gamete-producing phase in the life history.

4. (Gr. *gamete*, a wife; *phyton*, plant) the gamete-forming haploid phase in the alternation of plant generations.

- **gametophyte, gametophore**
  the part of the plant which bears the gametes or sexual cells. The haploid state of life of a sexually reproducing plant.

- **gamopetalous**
  1. having the petals joined at the base.
  2. having the sepals united. Also monosepalous.

- **gamophyllous**
  having united leaves or leaf-like organs. Also monopetalous.

- **garden compost**
  an organic matter such as well-decayed leaves, grass-clippings and vegetable waste, added to the soil as an amendment to improve its texture and drainage and to enrich it with nutrients.

- **garside's law/rule**
  the name given to an unusual arrangement in developmental tetrads of tri-aperturate pollen whereby the apertures form in groups of three at four points in the tetrad.

- **gazebo**
  an ornamental garden structure, resembling a free-standing porch, which may contain benches or statuary.

- **geitonogamy**
  the pollination of a pistil by pollen from another flower of the same plant.

- **gelatinous**
  a slimy and sticky jelly-like substance with a water base.

- **geminate**
  arranged in pairs.

- **geminicolpate (adj.)**
  describing pollen grains with colpi arranged in pairs. See **pontoperculum.**

- **geminiflorous**
  having flowers in pairs.

- **gemma**
  1. a vegetative reproductive bud borne on the stem, as in *Lycopodiu.* A multicellular reproductive propagule on gametophytes, as in ferns.
  2. (L. *gemma*, a swelling, bud, gem) a bud or outgrowth of a plant which develops into a new organism. A leaf bud rather than a flower bud.
  3. a small asexually-formed bud commonly produced by temperate Pinguicula, capable of

producing a plant genetically identical to the parent plant.
4. a bud, in some cryptogams, an asexual, bud-like reproductive body.
5. a sexine element which is constricted at its base, higher than 1μm and that has approximately the same width as its height.

- **gemmaceous**
relating to gemmae.

- **gemmate**
having gemmae, reproducing by buds.

- **gemmates**
(L. *gemmare*, to put forth buds) buds or outgrowths of a plant which develop into a new individual.

- **gemmation**
the act of budding; asexual reproduction in which a new organism originates locally within the body or on the surface of the parent. Also gemmulation.

- **gemmiform**
resembling a gemma.

- **gemmiparous**
(L. *gemma*, a bud; *parere*, to bear) to produce gemmates.

- **gemmoid**
having the nature or form of a gemma.

- **gemmule**
a small bud or gemma.

- **gene**
1. an inheritable factor, carried on the chromosomes, determining a character or trait in an organism.
2. the parts of the chromosomes of the cells of plants that determine inherited traits. For every trait or characteristic (i.e. flower colour), a plant usually receives one gene from each parent.
3. a segment of DNA comprising regulatory sequences and a coding region that specifies the amino acid sequence of a particular protein.

- **generative cell**
divides to form two sperm. The cell in a pollen grain which divides to form male gametes.

- **generic**
1. pertaining to a genus or a kind or group of similar things, comprehending a number of like things, without specifying them.
2. having the taxonomic rank or classificatory value of a genus (i.e. a generic name or description, generic characters or differences, generic identity).
3. distinctly characteristic, so marked as to constitute or denote a distinct kind.

- **generic agreement/identity**
the agreement of objects that belong to the same genus.

- **generic area**
the region to which the members of a genus (group) are limited in distribution. (The place in a generic area where the genus is most represented by species or individuals is known as its metropolis).

- **generic description/diagnosis**
a description or characte -risation of a genus in botany.

- **generic diversity**
the differences between individuals of different genera.

- **generic name**
the denomination that comprehends all the species, as of a group of animals, plants or fossils, which have generic characteristics in common.

- **genesis**
the means of development and origin of soil.

- **genetic engineering**
a process in which an isolated and well characterised gene is delivered, integrated and expressed in a host cell.

- **genetic map**
a figure demonstrating the relative position of specific genes along a chromosome.

- **genetic marker**
a small region of DNA along a chromosome that can be used to indicate the location of a gene or other DNA sequence.

- **genetic modification**
the alteration to an organism's genome by any number of methods, including inserting, transferring or deleting genes or other DNA sequences.

- **genetics**
the study of inheritance of characteristics.

- **geniculate**
bent at an abrupt angle, like a knee.

- **geniculum**
flexible portion of the thallus, such as the non-calcified segments between calcified intergenicula of articulated corallines.

- **geniculum (pl. genicula, adj. geniculate)**
a bulge in the equatorial exine of the colpus, often associated with a separation of the sexine from the nexine and the rupturing of the latter. Examples: Fraxinus (Olea ceae), Quercus (Faga- ceae).

- **genlisea (g)**
a genus of 15 species found in South and Central America and

in Africa, with an unusual 'corkscrew' trap unlike those found in any other genus; sometimes called Corkscrew Plant.

- **genome**
  the complete sequence of DNA, including all of the genes, of an organism.

- **genomics**
  the study of the genome.

- **genotype**
  1. the genetic characteristics of an organism.
  2. the genetic make-up of an organism. It usually describes the alleles present in a specific gene locus.
  3. (Gr. *genos*, race, type) the genetic constitution of an individual.

- **genus**
  1. a group of related species with sufficiently distinct unique characteristics in common to be treated as a separate unit. The name of the genus forms the first part of the binomial name of each species and is always capitalised.
  2. (L. *genus*, race) a taxonomic group consisting of closely related species, genera being grouped into families; plural- genera, generic.
  3. a category in plant classification, consisting of one or more related *species*. Thus, all Roses are grouped in the genus Rosa, all Willows in the genus Salix and all Oaks in the genus Quercus. Each genus or group, includes a number of species. For example, Quercus contains the Pin Oak (Q. palustris), the Red Oak (Q. coccinea), the Black Oak (Q. velutina) and many others. Some genera, such as Ginkgo and Sanguinaria, consist of only a single species. The genera name is also known as the botanical name. Just as there are species assigned to a single genus, there a number of genera assigned to a *family*.

- **geophyte**
  (Gr. *ge*, earth; *phyton*, plant) plants with an underground dormant part such as a tuber, bulb, rhizome etc. to help the plant survive adverse conditions.

- **geotropic**
  pertaining to, geotropism.

- **geotropism**
  a growth response to gravitational forces causing bending, growth or movement of a part under the influence of a gravitational field.

- **germ tube**
  a tubular, thread-like growth first formed by most spores in germination.

- **germinal aperture**
hole in the furrow membrane through which the pollen tube emerges. Examples: germ pore (synonym of pore), germinal furrow (synonym of ectocolpus). Synonym of exitus.

- **germinate**
to cause to grow.

- **germination**
1. the act, process or result of germinating ie.the development of a seed. The development of a seed into a plant. Oxygen, warmth and moisture are necessary for germination to occur. The moisture softens the seed, relaxes tissues in the embryo and dissolves nutrients. Oxygen is necessary for the chemical changes needed for growth. The temperature needed varies between the species. Some seeds, such as wheat seeds, will germinate on melting ice. Others need temperatures higher than 60o F. When germination takes place, a root is produced that grows down into the soil. At the same time, a stem and leaves are growing upward.
2. the similar development of a plant from a spore in cryptogams, such as ferns.
3. the early period of growth in a bud, as a bulb or a rhizome.
4. the protrusion and growth of the pollen tube from the pollen grain.

- **germinative**
capable of germination.

- **gibberelic acid (n)**
1. a plant hormone that stimulates the growth of shoots and leaves.
2. a plant hormone similar to gibberelic acid.

- **gibberellins**
a group of plant hormones based on gibberellic acid produced by seeds, mosses, ferns, algae and fungi. Gibberellins stimulate synthesis of nucleic acids and proteins and promote cell elongation, meristematic growth at the root and shoot tips, mobilisation of food reserves in seeds and other processes. Their absence results in the dwarfism of some plant varieties. These relatively expensive chemicals have been used commercially in the production of seedless grapes and navel oranges and in plant micropropagation from tissue culture and have been found to reverse the effects of some herbicides. Compare with auxins which inhibit root growth in high concentrations.

- **gibbous**
1. swollen on one side, usually

near the base, to form a pouch or sack.
2. (L. *gibbus*, hump) a distended, rounded swelling on one side, as on a calyx or corolla tube or segment.

- **gibrous**
swollen at one side.

- **girdle**
part of a diatom frustule between valves, also called cingulum.

- **glabrate**
(L. *glaber*, smooth) becoming glabrous with age.

- **glabrescent**
becoming glabrous.

- **glabrous**
smooth, without hairs or protrusions. Also glabrate.

- **glade**
(prob. Scand. *golead*, a lighting, illumination, fr. *goleu*, light, clear, AS. *glaed*, bright) open space surrounded by woods or a forest, a marshy and usually low-lying area. A periodically inundated grassy marsh often running between adjacent slopes, a marshy area bounding or forming the headwaters of a stream.

- **gladiate**
sword-shaped, ensiform.

- **gland**
1. a multi- cellular secretory structure producing oil, resin or other liquid, usually as a protruberance or appendage, but sometimes as part of a surface.
2. any secreting organs of plants or of gland-like bodies.
3. (L. *glands*, acorn) a secreting part or appendage.
4. a secreting cell or group of cells on the surface of a plant structure.

- **gland cell**
special, refractive cell in some red algae which may function in secretion or storage of compounds.

- **glandular**
1. resembling or pertaining to a gland. Bearing glands. Derived from glands. Also glandulous.
2. (L. *glandula*, small acorn) having or bearing secreting organs, glands or trichomes.
3. pertaining to the nature or action of glands.

- **glandular-pubescent**
hairs or trichomes capitate and secretory.

- **glanduliferous**
capable of or engaged in secretion as a gland.

- **glans**
a nut seated in or enclosed by an involucre.

- **glaucescence**
  a fine bloom, blush or coating of whitish or bluish-white substances, usually microscopic particles of wax, easily rubbed off the surface.
- **glaucescent**
  slightly glaucous.
- **glaucous**
  bluish grey or bluish green often as a result of a wax-like bloom. Having a waxy or powdery, bluish-white or bluish-gray bloom that can be rubbed off or covered with a pale green bloom. Covered in powdery bloom - as of grapes.
- **gley soil**
  a layer of soil that has turned gray because of lack of oxygen and excessive water.
- **gliding movement**
  movement of an organism when in contact with substratum.
- **globose**
  spherical, spheroidal, rounded, almost spherical or globular.
- **glochidiate**
  barbed at the tip. Also glochideous.
- **glochids**
  spines or hairs, usually small, brittle and barbed. Occurring in tufts and characteristic of Opuntioideae. Barbed bristly hairs or spines, often very fine and easily detached, on an Opuntia cactus.
- **glomerate**
  1. compactly clustered.
  2. vt. (L. *glomus*, a ball of yarn) to gather or wind into a ball-growing, collected or arranged in a rounded mass, as glands, flowers etc., clustered.
  3. clustered compactly.
  4. arranged in small, compact clusters. Also glomerulose.
- **glomerule**
  1. a compact flower cluster that forms a head, as in the flowering Dogwood, Cornus florida.
  2. a condensed cyme of almost sessile flowers. A compact cluster as of spores.
- **glumaceous**
  glume-like, having glumes, chaffy.
- **glume**
  1. a chaffy bract as in the Gramineae (grass family) and related plants, especially an empty bract at the base of a grass spikelet.
  2. (L. *gluma*, husk) a chaffy or membranous bract, a bract at the

base of a grass inflorescence or spikelet.

- **glumiferous**
having glumes.

- **glutinous**
(L. *gluten*, glue) having a sticky, moist surface. A gluey or sticky exudation.

- **glycoprotein**
a diverse group of proteins which are modified by addition of short chains of typically 8-10 sugar residues covalently linked to their polypeptide structure.

- **GM (genetically modified) foods**
foods modified by genetic manipulation (a type of novel food).

- **gongyloid**
irregularly round in appearance.

- **gonimoblast**
diploid filaments of carposporophyte (formed in Rhodophyta after fertilisation) that bear the carpospores.

- **goniotreme**
synonym of angulaperturate.

- **gracilis**
slender.

- **graft**
1. a shoot, bud or other tissue joined to and growing on, another compatible plant.
2. method of propagation by which an artificial union is made between different parts of individual plants. The stem or shoot that is inserted into a rooted plant is called the scion, the plant or part of a plant into which the scion is inserted is called the stock or understock. There are many different methods of grafting including flat grafting, split grafting and side grafting, among others.
3. the point where a scion is inserted in the stock.

- **graft hybrid**
a hybrid produced by grafting a branch of one species (scion) onto another (rootstock), in which the two grow together to form a single plant or tree.

- **grafting tape**
tape backed with biodegradable cloth. Used in budding and grafting operations and in band-

ing tree wounds. Also called nurseryman's tape.

- **grafting thread**
  fine waxed string used in budding and grafting operations.

- **grafting wax**
  a waxy substance used to hold and seal tree or rose grafts.

- **grain**
  the unhusked or threshed seed of one of the cereal plants: wheat, oats, rye, barley, maize or millet. Corn in general, the gathered seeds of cereal plants in mass.

- **gramineous**
  grass-like, belonging to the family Gramineae. Also graminaceous, gramineal.

- **graniform**
  having the form of a seed or grain.

- **granular**
  containing, bearing, resembling or composed of grains or granules. Covered with small grains or granules. Also granulose, granulous.

- **granular exine**
  a type of exine stratification in which the infratectal layer is composed of more or less rounded, granules rather than of columellae or other structures. See **columella, alveolate.**

- **granular fertiliser**
  a fertiliser made up to particles 0.1 inch in diameter.

- **granulate**
  to form into granules, make rough on the surface. To become formed into grains or become granular, consisting of or resembling grains or granules. Also granulated.

- **granulate granulatus**
  with granules.

- **granulation**
  the process of granulating. One of the little elevations in a granulated surface.

- **granule**
  a little grain, a fine particle. General word for a small, rounded element.

- **granulum (pl.granula, adj.granulate, granulose)**
  a very small and rounded element of the sexine/ectexine that is less than 1μm in all directions.

- **granum (pl. grana)**
  the combined elements gemmae and verrucae.

- **gravitational water**
  the water, in large pores of the soil, which drains away under the force of gravity.

- **great soil group**
any one of several broad groups of soil with fundamental characteristics in common.

- **green manure**
an organic crop material, turned under the soil, to be used as a fertiliser to add nitrogen to the soil. The green plants decay in the soil and provide humus and make stiff soil more porous and friable. A quick-growing crop is sown and dug into the ground before it reaches maturity (when it will provide the greatest bulk without being to tall to turn under).

- **green seed**
a canola production problem in which an early frost inhibits the degradation of the green pigment chlorophyll. The presence of chlorophyll in the oil lowers the yield and grade of oil obtained from the seed.

- **greenhouse**
a building of glass or plastic that is used to grow and protect tender plants.

- **greenhouse effect**
the trapping of heat from sunlight by atmospheric gases (greenhouse gases), including carbon dioxide, methane and water, which raises planetary temperatures above those expected from a simple model of heat radiation received from the sun and re-radiated into space. The high surface temperature on the planet Venus is caused by an extreme form of greenhouse effect.

- **ground cover**
refers to low-growing or trailing plants that are used to carpet the ground, as underneath trees or as an overplanting for bulbs.

- **ground meristem**
develops into the ground tissue system.

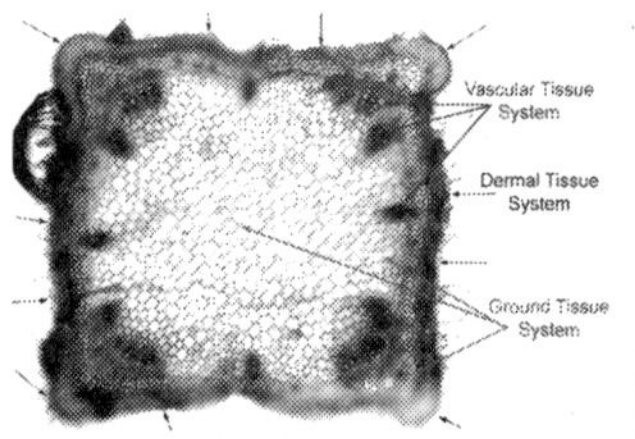

- **ground tissue**
a tissue consisting mostly of parenchyma cells that makes up the bulk of a young plant.

- **ground tissue system**
tissues derived from the ground meristem. All are simple tissues composed of a single type of cell, which is named after the tissue.

- **ground water**
water that fills all unblocked pores of underlying material

below the water table. The upper limit of saturation.

- **grow light**
a broad-spectrum florescent light used to grow plants indoors.

- **growing point**
the tip of a shoot from which new growth emerges.

- **growing tray**
a tray having compartments like an ice-cube tray, used for starting seeds. Also called a cell tray.

- **growth form**
general description of the type of growth exhibited by a plant, such as herbaceous, shrubby (bush-like) and arborescent (tree-like).

- **growth form**
the form of a plant, the habit in which a plant grows also called growth habit or just habit. Shrub and vine are growth forms (i.e. a shrubby plant or a climbing plant - a vine).

- **growth ring**
the layer of wood developed in a tree's growing season, usually one year.

- **grumous**
in the form of clustered grains, coarse grains or granules. Also grumous.

- **guano**
a strong organic fertiliser composed of bird droppings.

- **guard cell**
one of two crescent-shaped cells which surround and control the opening and closing of the aperture of the stoma and thereby regulate passage of gases and water vapour through a plant's epidermis. Cells that surround and control the size of stomatal pores.

- **gula (pl. gulae, adj. gulate)**
a rather ornate projecting, neck-like, extension on the proximal face of a trilete spore. example: Lagenicula horrida.

- **gum**
material resulting from the breakdown of plant cells including their carbohydrates. On exposure to air, gums dry and harden.

- **gummifera**
bearing gum.

- **guttate**
spotted, as by drops. Also guttated.

- **guttation**
1. exudation of liquid water from plants.
2. (L. *gutta*, drop) formation of drops of water on plants from moisture in air, the process of

water being exuded from hydathodes at the enlarged terminations of veins around the margins of the leaves.

- **guttered**
  of leaf margins folded inward.

- **guttiferous**
  producing gum or resinous substances.

- **gymnocarpous**
  having a naked fruit, especially in lichens.

- **gymnogynous**
  1. having a naked ovary.
  2. with naked or unprotected ovules (seeds) e.g. conifers, cone-bearing plants such as *Welwitschia*. Gymnosperms do not produce flowers, but instead release pollen through the air to the female ovule, causing fertilisation.
  3. a plant belonging to the Gymnospermae, characterised by naked seeds, as in the conifers (plants producing cones).

- **gymnospermae**
  (Gr. *gymnos*, uncovered, naked; *sperma*, seed) an important division of the plant kingdom, being woody plants with alternation of generations, having the gametophyte retained on the sporophyte and seeds produced on the surface of the sporophylls and not enclosed in an ovary.

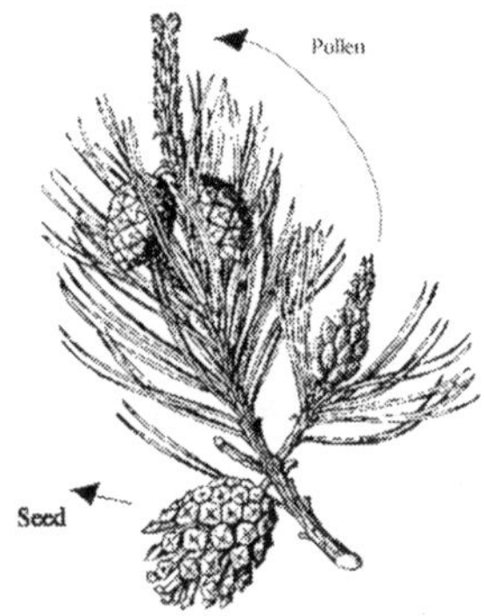

- **gymnospermous**
  having naked seeds. Also gymnogenous, gymnospermal, gymnospermic.

- **gymnospore**
  a naked spore.

- **gymnosporous**
  having naked spores.

- **gynaecium**
  (Gr. *gynaikeie*, woman's part of a house) the female organs of the flower, consisting of one or more carpels forming one or several ovaries with their stigmas and styles.

- **gynandrous**
  having the stamens *adnate* to and apparently borne upon the pistils, as in all Orchids.

- **gynecandrous**
  having staminate and pistillate flowers in the same spike or

spikelet, the latter above the former.

- **gyno-**
pistil or ovary.

- **gynobase**
a prolongation or enlargement of the flower receptacle that bears the gynoecium.

- **gynodioecious**
having perfect and female flowers growing upon separate plants.

- **gynoecium**
1. a collective term for all female organs of a flower. The pistil or collective pistils of a flower, the female portion of a flower as a whole. In correlation to *androecium*. Also gynecium, gynaeceum.
2. (Gr. *gyne*, woman; *oikos*, house) the pistil or pistils of a flower, taken collectively; gynaecium.
3. the carpels of a flower, collectively.

- **gynomonoecious**
having both perfect and female flowers upon the same plant.

- **gynophore**
an elongation of the receptacle of a flower, bearing the gynoecium at its apex, as in some Capparidaceae.

- **gynophore**
(Gr. *gyne*, woman; *pherein*, to carry) a stalk supporting the ovary.

- **gynostegium**
1. a covering of the gynoecium.
2. (Gr. *gyne*, woman; *stege*, roof) a protective covering for a gynaecium, especially as formed by the union of stamens and style.

- **gynostemium (pl. gynostemia)**
the column of an orchid, formed by the union of the *androecium* and gynoecium.

- **gyroma**
the annulus of a fern. The shield of some lichens.

- **habit**
1. the general form and growth pattern of a plant.
2. the characteristic growth or general appearance of a plant. The 'habit of growth' may be low-growing, tall, erect, spreading, trailing, bushy and so on.

3. (L. *habitus*, condition, appearance, dress) the external appearance or way of growth of a plant, e.g. climbing, erect, bushy etc. the tendency of a plant to grow in a certain way.

- **habitat**
  the place in which a plant or other organisms, naturally grows or lives.

- **hair**
  a long thin epidermal appendage, either formed from a single elongated cell or consisting of a single row of cells.

- **hairs**
  colourless, elongate unicellular or multicellular structures.

- **half life**
  the time taken for the value of something to reduce to half of that value. Of a radioactive isotope, the time taken for half of the atoms to decay.

- **half-hardy annuals**
  plants that will survive some frost, but not a long freeze.

- **halo**
  a clear zone around a well defined feature such as a spine or an aperture. Examples: Ranunculus acris (Ranunculaceae), Valeriana (Valerianaceae).

- **halomorphic**
  characters that are advantageous in saline environments.

- **halophilous**
  preferring or habitually growing in salty soil as maritime plants.

- **halophyte**
  1. a plant adapted to, growing in and tolerating salty environments.
  2. any species capable of tolerating 0.5% or more NaCl.

- **hamate**
  curved at the tip, like a hook.

- **hamulate**
  describing a form of rugulate ornamentation consisting of irregularly arranged, winding or angular rounded muri of varying thickness, which do not form a distinct reticulum, but rather a maze-like pattern. Example: Lycopodiella inundata (Lycopodiaceae).

- **handpicking**
  a method of controlling pests by literally removing the pest from the plant by hand. This is effective with larger pests, such as tomato hornworms and snails.

- **hanging basket**
  a container hung from a brace, tree etc., with a rope or cord,

which is used to grow ornamental plants.

- **hanging plant rotator**
a device that rotates a hanging plant a certain amount (usually 90°) to ensure even lighting.

- **haplobiontic**
having single, free-living phase in its life history.

- **haploid**
having the number of chromosomes characteristic of the germ cells or half the number present in the body cells.

- **haplophyte**
(Gr. *haploos*, simple; *eidos*, form) having the number of chromosomes characteristic of the gametes for the organism.

- **haploxylonoid**
describing bisaccate pollen in which the outline of the sacci in polar view is more or less continuous with the outline of the corpus, so that the grains appear a more or less smooth ellipsoidal form. See **diploxylonoid, Haploxylon-type, Sylvestris-type.**

- **haploxylon-type**
bisaccate pollen in which the outline of the sacci in polar view is more or less continuous with the outline of the corpus, so that the grains appear a more or less smooth ellipsoidal form. Examples: Pinus cembra, Picea (Pinaceae). See **diploxylonoid, haploxylonoid, Sylvestris-type.**

- **haptera**
multicellular structures that form holdfasts of kelps.

- **hapteron**
(Gr. *haptein*, to fasten) holdfast, specialised root-like projections that function to anchor a plant.

- **haptonastic**
a movement response to a touch stimulus.

- **haptonema**
cellular extension with distinctive structure characteristic of some Prymnesiophyceae.

- **hardening off**
the gradual process of acclimating plants started indoors to outside conditions. Setting them in a sheltered location outdoors for increasing lengths of time over a period of days is said to be hardening off the plants. If the plants are just set outside without getting used to the different atmospheres, they will most likely suffer.

- **hardpan**
a hard or cement-like soil or soil layer.

- **hardwood**

strong, dense wood from slow-growing broad leaved trees. Thewood from trees with encapsulated seeds (i.e. nuts or fruits), as distinguished from a conifer or cone-bearing tree, which has naked seeds. Oaks, Walnuts and Maples are all hardwoods. Pines and firs are softwoods.

- **hardy**

used to describe a plant that is able to survive the cold temperatures of a given location without protection. When a range is given, hardiness also describes a plant's ability to tolerate the hottest temperatures normally experienced. (For example, a Palm tree isn't hardy in Alaska, because it is a tropical plant requiring a warm, humid atmosphere.)

- **harmomegathy**

the process by which pollen grains and spores change in shape to accommodate variations in the volume of the cytoplasm caused by changing hydration.

- **harrow**

a rake-like tool that is drug over plowed soil to prepare it for planting. To draw a harrow over soil, to break or tear with a harrow, as to harrow ground.

- **hastate**

1. having a triangular shape like a arrowhead.
2. (L. *hasta*, spear) spear shaped, more or less triangular with the two basal lobes divergent.
3. a triangle with two sides abruptly concave; in the form of two pointed lobes pointing outwards.

- **haulm**

the collective stems or tops of crop plants (e.g. beans, peas, potatoes, grains, grasses) after the crop has been gathered, as used for bedding or thatching. Occasionally applied to a single stem.

- **haustaurium**

(Pl. haustoria)- a specialised organ through which a parasitic plant absorbs nourishment from its host.

- **haustorial root**

a root that is modified for absorbing water or nutrients from another plant.

- **head**
  the top of a plant consisting of a compact mass of leaves, such as a head of lettuce. A short, dense bunch of flowers. To cut back the shoots of plants in order to induce lateral growth.

- **heartwood**
  the older, non-living central wood in a tree.

- **heat mat**
  a low-voltage heater used to warm soil during seed germination, seedling development and root growth of cuttings.

- **heath**
  open land covered with low-growing shrubs such as heather.

- **heating cable**
  an electric device used to heat soil in a greenhouse or cold frame.

- **heating system**
  a unit consisting of an adjustable temperature controller and a heat mat. Used to maintain soil temperature during germination and seedling development or during root growth of cuttings.

- **heaving**
  the pushing of plants out of the soil because of alternate freezing and thawing action.

- **heavy soil**
  fine-textured, clay soil.

- **hebetate**
  a soft, blunt termination.

- **hectare**
  (ha) a metric unit of area equal to 100 Ares (2.471 acres) and equivalent to 10,000 square meters (107,639 square feet).

- **hedgea**
  a barrier or fence formed by bushes or small trees growing together, a closely planted row of shrubbery. To enclose or fence with a hedge, to separate by a hedge.

- **hedge - hedgerow**
  a border, fence or boundary formed by a row of closely planted shrubs or bushes often trimmed to a formal shape. The number of species forming a hedge tends to increase in proportion to its age.

- **hedge binding**
  something that is used to bind together the bushes composing a hedge.

- **heel**
  1. a small portion of the previous year's growth at the base of a cutting.

2. when a cutting it taken with a 'heel' of wood, it means a piece of the old branch or shoot is detached from the old branch or shoot along with the cutting.

- **heeling in**
  temporarily covering the base of a plant, especially a dormant plant, with soil for a short time.

- **heliamphora (g)**
  a genus of five species of pitcher plants found in South America.

- **helicoid**
  tending to a helical shape or pattern.

- **heliophilous**
  fond of the sun, attracted by or becoming most active in sunshine.

- **heliophobic**
  shade-loving.

- **heliotropic**
  turning toward the sun, as leaves or flowers.

- **heliotropism**
  the tendency of plant parts, such as foliage, stems or flowers, to position themselves according to the direction in which the sun hits them.

- **helobious**
  living in swamps or marshes.

- **hemicellulose**
  the alkali-soluble portion of the cell wall matrix.

- **hemicryptophyte**
  (Gr. *hemi*, half; *kryptos*, hidden; *phyton*, plant) a perennial plant having its overwintering buds located at the soil surface.

- **h-endoaperture**
  an elaborate endoaperture, consisting of a central part which connects two lateral, longitudinal elongations, forming an 'H' shape. Examples: Cornus (Cornacaea), Centaurium (Gentianaceae).

- **henequen**
  a strong fibre, similar to sisal, which is produced from the leaves of a Central American Agave and used mainly for making twine. Also spelled heniquen.

- **herb**
  1. a vascular plant lacking a woody structure.
  2. a plant grown for its medicinal or flavoring properties or for its scented foliage.
  3. also refers to plants without persistent, woody stems. This botanical use of the word 'herb' is used to distinguish between trees and shrubs.
  4. (L. *herba*, green crop) any seed plant whose stem withers away

to the ground after each season's growth; a seed plant with a green, non-woody stem.

- **herbaceous**
  dying to the ground periodically or lacking a definite woody structure. A non-woody plant in which the upper parts die back to the rootstock at the end of the growing season. Example-Peony.

- **herbage**
  1. leafy vegetation.
  2. the herbaceous parts of a plant.
  3. (Fr. *herbe*, an herb) herbs collectively.
  4. the green foliage and juicy stems of herbs.

- **herbal**
  pertaining to or consisting of herbs.

- **herbalism**
  the knowledge of herbs.

- **herbalist**
  someone who is skilled in the knowledge of plants or makes collections of them. A dealer in medicinal plants, a person who treats diseases with botanical remedies only.

- **herbarium**
  a collection of dried plants systematically arranged. The plants are attached to sheets of paper, either by small gummed strips of paper or by gluing one side of the specimen. A systematically arranged collection of dried plants or record of such.

- **herbivore**
  an animal that eats only plants.

- **herborist**
  an herbalist.

- **herborisation**
  an excursion for studying or collecting plants.

- **hercogamous**
  incapable of self-pollination.

- **hercogamy**
  a condition in which structural obstacles within a flower make self-pollination impossible.

- **hermaphrodite**
  with both male and female reproductive parts usually in the same flower. Bisexual, bearing both male and female organs in the same flower, having fully

developed stamens and pistils in the same flower.

- **hesperidium**
  a fruit with a leathery rind (i.e. an orange).

- **heterauxesis**
  the irregular growth of a plant due to unequal rates of growth in its parts.

- **heteroblastic leaves**
  leaves produced during one stage of development are distinct from those formed at another stage, for example leaf from may change when a plant flowers.

- **heterobrochate**
  describing a reticulum with brochi of different sizes. Examples: Adoxa Moschatelina (Adoxaceae), Forsythia europaea (Oleaceae). See **brochus, lumen.**

- **heterocellular rays**
  rays consisting of more that one type of cells.

- **heterochrony**
  an evolutionary change in phenotype based on an alteration in timing of developmental events.

- **heterocolpate**
  describing pollen grains with both simple and compound colpi present. Examples: Lythrum, Peplis (Lythraceae), Myosotis Boraginaceae). See **pseudocolpus.**

- **heterocyst**
  specialiased, thick-walled cell of some Cyanobacteria which may be site of nitrogen fixation.

- **heterodromous**
  having the genetic spiral of the branches different from that of the main stem, affecting the way in which leaves are arranged around a stem.

- **heterodromy**
  a difference in direction of the genetic spiral in the branch and the main stem.

- **heterogamous**
  sexual reproduction where gametes are not identical in size or shape. Having 2 or more distinctly different forms of flower.

- **heterogamy**
  the condition of being heterogamous.

- **heteromerous**
  in reference to flowers, having the parts of adjoining whorls different in number.

- **heteromorphic**
  having irregular, abnormal or unusual structure, as the leaves of a young plant. Having life history phases morphologically different, having distinctly dif-

ferent gametophyte and sporophyte stages.

- **heteromorphism**
  the condition of being heteromorphic.

- **heteromorphus**
  with an irregular or unusual shape or structure.

- **heterophyllous**
  (Gr. *heteros*, other; *phyllon*, leaf) the presence on a single individual of two or more distinct leaf shapes. These leaves may differ markedly in shape, yet have similar gross anatomical organization. Producing different leaves in Winter and Summer.

- **heterophylly**
  the presence of more than one type of leaf on the same plant, such as floating or aerial leaves and submersed leaves on the same aquatic plant.

- **heteropolar**
  describing pollen or spores in which the distal and proximal faces of the exine are different, either in shape ornamentation or apertural system. Example: Echium vulgare (Boraginaceae).

- **heterosporous**
  1. having two types of spores: megaspores and microspores.
  2. having more than one kind of asexually produced spores (applied to ferns).
  3. having two kinds of spores, usually differing in size.
  4. describing plants producing both microspores and megaspores.

- **heterospory**
  a condition in which an organism produces two different types and sizes of spores, viz. microspores and megaspores.

- **heterostyled**
  having the styles of the flower of two or more different forms.

- **heterotactic**
  characterised by heterotaxis.

- **heterotaxis**
  abnormal arrangement of parts or organs as compared to a normal type.

- **heterotrichous (filament)**
  a growth pattern of algae with prostrate filaments for attachment and erect, spreading filaments for photosynthesis.

- **heterotrichy**
  filamentous thallus having erect and prostrate parts.

- **heterotroph**
  an organism that can not make its own food and consumes other organisms to survive.

- **heterotrophic**
having non-autrotrophic, non-photosynthetic nutrition.

- **heterozygote**
1. with unlike genes on any particular pair of chromosomes.
2. (Gr. *heteros*, other; *zygon*, yoke) an organism or cell having two different alleles at corresponding loci on homologous chromosomes.

- **hexapetaloid**
of a flower, having six petals. Also hexapetalous.

- **hibernaculum**
1. the part of a plant that is dormant in the winter, as a bulb or bud.
2. (L. *hibernare*, winter residence) a plant organ such as a bud, rhizome, turion etc. which allows a plant to live through adverse conditions.
3. a resting bud formed by many temperate and boreal Drosera and Pinguicula as an aid to surviving the Winter months.

- **hibernation**
spending the winter months in a resting or dormant state.

- **hilar**
belonging to the hilum or scar, on a seed produced by the attachment to the pod, as with beans.

- **hilum**
the small scar on a seed coat where it was attached to the seed pod by the funicle as with beans or peas. Circular, indistinctly delimited, irregular aperture or thinning in spores. Examples: Aequitriradites verrucosus, Couperisporites tabulatus.

- **hirsute**
1. pubescent, with coarse, stiff hairs.
2. having coarse, rigid hairs.
3. (L. *hirsutus*, bristly) set with bristles, hairy, shaggy.

- **hispid**
having stiff, bristly hairs or tiny spines.

- **hispidity**
the state of being hispid.

- **hispidulous**
somewhat hispid.

- **hoary**
covered with a whitish or grayish-white down.

- **holdfast**
1. algal organ for attachment.
2. structure of attachment to substratum.

- **holistic**
of, concerned with or dealing with whole or integrated systems rather than with their parts.

- **holoplankton**
  organisms that spend their entire existence as free-floating drifters.

- **holotype**
  the type specimen of a taxon as designated in the author's description.

- **homobrochate**
  describing a reticulum with brochi of the same sizes. Example: Armeria maritima (Plumbaginaceae).

- **homocellular rays**
  rays consisting on one type of cell.

- **homogamous**
  of a plant, having all the flowers alike. Of a flower, having the stamens and pistils maturing at the same period.

- **homogamy**
  the state of being homogamous.

- **homogonous**
  having the stamens and pistils of the same height in members of the same species.

- **homogony**
  the state of being homogonous.

- **homologous, homology**
  homology: likeness and correspondence in structure between parts of different organisms, due to common ancestry of the organisms, compare to analogy.

- **homology**
  having the same phylogenetic origin, but not necessarily identical form or function.

- **homomallous**
  referring to the leaves of mosses, curving to one side.

- **homonym**
  a name that already exists for another plant. The newest name is illegitimate and must be changed.

- **homopetalous**
  having all the petals formed alike.

- **homophyllous**
  producing the same form of leaf in the Winter and Summer months.

- **homoplasy**
  of different species with similar characters that have evolved separately and are not from a common phylogenetic origin. Same as convergence.

- **homosporous**
  having one type of spore.

- **homospory**
  a condition in which an organism produces only one type and size of spore, viz. microspores.

- **homostyled**
  having all styles one length.
- **homozygote**
  with identical genes on any particular pair of chromosomes.
- **honeydew**
  1. a sweet substance found on the leaves of trees and other plants in small drops, like dew. It is the sugary substance secreted from the leaves of plants, occurring most often in hot weather. It usually is seen as small, glistening drops, but it may drip abundantly from the leaves, when it has been called manna. This especially happens with Fraxinus Ornus (the Manna Ash) and Carduus arctiodes.
  2. a sweetish secretion exuded by many insects, such as aphids, mealybugs, scale and whiteflies. The secretion attracts ants, which harvest the honeydew and protect the pests from natural predators. It also develops a blackish mold, known as sooty mold. Plants that are heavily infested, especially trees, drip honeydew, staining anything underneath, such as vehicles. The best control is to eliminate ants (banding with sticky materials is effective) from the plant and then control the source pest. Insecticidal soaps will help clean the foliage.
- **hook climber**
  a plant that climbs by the aid of hooks or prickles, such as certain types of roses.

- **hooked**
  with a straight basal portion and a tip curved into an arc.
- **horizon soil**
  a layer of soil approximately parallel to the soil surface.
- **horizonthalonius**
  with a flat, level disc.
- **hormone**
  1. a chemical substance produced in a plant tissue, capable of diffusing or transport to other parts where it's presence or gradients regulates specific functions such as abscission, differentiation, flowering, growth and tropisms.
  2. any of certain chemical substances that are produced by living cells in one part of the plant and initiate cellular activity (such as growth or flowering) in a distant part of the plant.

3. a chemical produced by organisms to trigger or regulate certain processes.

- **horn**
  an elongated part of an endoaperture, which is curved towards one of the poles. Example: Anagallis arvensis (Primulaceae).

- **horticultural**
  pertaining to horticulture.

- **horticultural oil**
  a refined oil sprayed on plants to control a variety of pests. It kills by smothering the pests and their eggs or by disrupting membrane functions. It may also prevent some fungal diseases. Highly refined types are called summer, superior or verdant oils and can be used during the growing season as long as temperatures are below 90° F. Heavier types are called dormant oils and are used on dormant, deciduous plants.

- **horticulture**
  1. the science of growing plants for decorative effect or for consumption as food.
  2. the cultivation of a garden, an orchard or a nursery; the art and science of cultivating flowers, fruits, vegetables and ornamental plants.

- **horticulturist**
  a person who specialises in horticulture.

- **hortus siccus**
  a collection of specimens of plants carefully dried and preserved for botanical purposes and herbarium.

- **hose guides**
  devices, anchored by stakes, used to route a garden hose above or around plants. The notched or open-loop-top, type is used to route a hose above or around plants. The spool type is used to guide a hose around corners.

- **hose-in-hose**
  a flower that appears to have a double corolla, one within another, as in certain Azaleas.

- **host**
  the organism on or in which a parasite lives and from which it derives its nourishment. Sometimes also used for the plant on which an epiphyte lives. A plant that supports a parasite that lives in or upon it.

- **hotbed**
  a bed of soil heated by fermenting substances and covered with glass to keep out cold air, intended for raising early plants or for protecting tender exotics.

- **hotcap**
  an individual, conical plant cover used to protect against frost, sun, birds and insects.

- **hothouse**
  a structure kept heated for the growth of tender, exotic plants or subtropical plants or for the production of native fruits, flowers etc., out of season. The temperature of the hothouse is kept between the temperatures of the greenhouse and the stove-house or orchid-house.

- **houseplant**
  any plant that is suitable for growing in the conditions of a house.

- **hull**
  the outer covering of a seed, nut or fruit.

- **humid acids**
  the term sometimes interchangeable for humus.The alkali-soluble end products of the decomposition or organic matter in mineral soils.

- **humidity meter**
  an instrument used to measure the amount of moisture in house or greenhouse air.

- **humming line**
  a ribbon suspended over crops that emits am ultrasonic sound to discourage birds.

- **humus**
  1. the organic component of soil, composed of partially or completely decayed plant and animal remains that provides nutrients for plants and increases the ability of soil to retain water.
  2. the organic residue of decayed vegetable matter in soil. Also often used to describe partly decayed matter such as leafmold or compost. Humus retains moisture and keeps the ground from drying out quickly in hot weather; it is the main indirect food supply for plants. Soil which lacks humus soon becomes impoverished. Humus is supplied mainly by adding animal manure, leaf mold, decayed garden refuse (compost), peat moss, sedge peat and by turning under crops grown as cover crops or green manures.

- **husk**
  the outer covering of the fruit of a plant (i.e. corn husks).

- **hyaline**
  1. translucent.
  2. (Gr. *hyalos*, glass) of thin, membranous, transparent or translucent texture.

- **hybrid**
  1. progeny from an inter-specific (or inter-generic) fertilisation, often of intermediate appearance to the parents.
  2. the offspring of genetically different parents, often produced accidentally or artificially in cultivation, but occasionally arising in the wild.
  3. a plant formed by the breeding of two different plant species.
  4. (L. *hibrida*, cross) any crossbred plant; heterozygote.
  5. a specimen produced by the sexual reproduction of 2 plants of different species or subspecies.

- **hybridisable**
  capable of producing hybrid offspring by crossing with another variety, species or genus.

- **hybridisation**
  the practice of crossbreeding between two species or wild types. This term is used to indicate crossbreeding generally, whether between species or varieties.

- **hybridise**
  to produce or cause to produce a hybrid.

- **hybridiser**
  a person who hybridises plants.

- **hydathode**
  a specialised gland or pore on the surface or leaf margin of higher plants that exudes water from the plant tissues.

- **hydathode**
  a structure in the margins of leaves that secretes water.

- **hydathodes**
  (Gr. *hydatos*, of water; *hodos*, way) an epidermal structure specialised for secretion or for exudation, of water.

- **hydric**
  (Gr. *hydor*, water) characterised by an abundant supply of water.

- **hydrogen**
  an element, found in the soil, that plants need in order to grow.

- **hydroid**
  the water conducting cells of bryophytes; compare to leptoid.

- **hydromorphic**
  characters that are advantageous in wet environments.

- **hydrophily**
  the flowers are produced under the surface of the water and that

is where the transfer of pollen takes place. Fairly rare occurrence.

- **hydrophyte**
  1. a plant that grows in and is adapted to an aquatic or very wet habitat.
  2. a plant with hydromorphic features.
  3. (Gr. *hydor*, water; *phyton*, plant) an aquatic plant living on or in water.

- **hydroponics**
  a method of growing plants in water rather than soil. The plant's roots are suspended in a nutrient solution.

- **hydropote**
  (Gr. *hydropotes*, water drinker) a cell or cell group found on the lower epidermis of some species such as *Nymphaea*. These cells are thought to function in the uptake of ions from the water.

- **hydrotropism**
  bending, growth or movement of a part in response to a water source.

- **hydrous**
  containing water.

- **hydroxyapatite**
  a member of the apatite groups or mineral rich in the hydroxyl groups.

- **hygrophyte**
  a plant growing in wet places.

- **hygroscopic movement**
  movement by the gain or loss of water

- **hypanthial**
  pertaining to or resembling a hypanthium.

- **hypanthium**
  1. the part of certain flowers usually formed by the fusion of the lower portions of their petals, sepals and stamens.
  2. a cup-shaped extension of the receptacle.
  3. (Gr. *hypo*, under; *anthodes*, like flowers) an expansion of the receptacle of a flower that forms a saucer-shaped, cup-shaped or tubular structure (often simulating a calyx tube) bearing the perianth and stamens at or near its rim; it may be free from or united to the ovary.

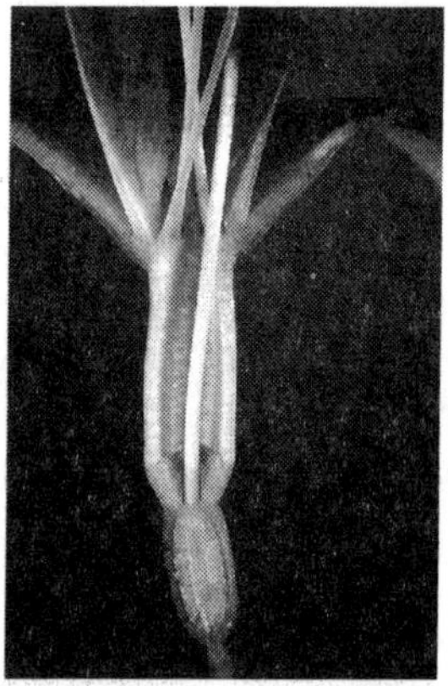

- **hyperplasia**
  an excessive multiplication of cells.

- **hypertrophy**
  excessive growth or overdevelopement of the organs of plants.

- **hypochile**
  the basal part of the *labellum* (of an Orchid) in some groups, such as *Stanhopea*.

- **hypocingulum**
  part of the girdle of a diatom frustule adjacent to the hypovalve.

- **hypocone**
  portion of dinoflagellate cell posterior to the cingulum.

- **hypocotyl**
  1. the main axis of the seed embryo just below the cotyledons and continuing into the developing primary root of the seedling.
  2. the part of the stem of a plant embryo or seedling that is below the seed leaves.

- **hypocotyledonary**
  located below the seed leaves.

- **hypocrateriform**
  salver-shaped

- **hypodermis**
  1. a layer of cells below the epidermis.

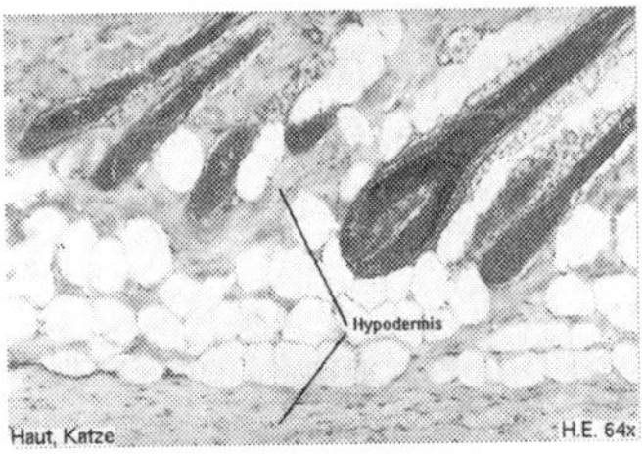

  2. a layer or layers of cells beneath the epidermis that is derived from the ground meristem, but distinct in appearance from adjacent ground tissue. May be called an *endodermis* if it has a Casparian strip.

- **hypogeal**
  happening, living or remaining below ground. Of the emergence of cotyledons below the surface of the ground or within the sead coat.

- **hypogenous**
  growing on the under surface, especially pertaining to a fungus on the underside of a leaf.

- **hypogeous**
  below ground. Having the sepals, stamens and petals arising from the portion of a flower situated below the ovary.

- **hypogynous**
  sepals, petals and stamens attached to the receptacle below the ovary.

- **hypogynous cell**
  cell directly beneath the carpogonium in the Rhodophyceae.
- **hypogyny**
  the state of being hypogynous.
- **hypolithic**
  living on lower surface of rocks.
- **hypophyllous**
  situated on the underside of a leaf.
- **hypostomatic**
  stomata in the lower epidermis only.
- **hypotheca**
  part of diatom frustule consisting of hypovalve and hypocingulum.
- **hypovalve**
  flattened or covex plate of diatom frustule opposite the epivalve.
- **hypsophyll**
  a leaf located underneath spore-bearing leaves.
- **hypsophyllary**
  pertaining to the hypsophyll.
- **IAA**
  Indole-Acetic Acid, a natural growth hormone found in plants.
- **icterus**
  jaundice.
- **idioblast**
  a special cell in a tissue which differs markedly in form, size or contents from other cells in the same tissue.
- **idioblasts**
  (Gr. *idios*, one's own; *blastos*, a bud, offshoot) plant cells containing oil, gum, calcium or other products and appearing to help provide mechanical support.
- **illuvial horizon**
  an accumulation of material in the B horizon (subsoil) that moved into it from the A horizon (topsoil).
- **imbricate**
  1. with overlapping edges.
  2. overlapping in regular order, as the scales on a snake.
  3. (L. *imbricare*, to tile) having parts overlapping each other like roof tiles.
  4. overlapping in a tile-like manner.
- **imc (isolated microspore culture) systems**
  offer an unlimited supply of embryogenic cells and developing embryos, which facilitate studies of key developmental processes revelant to seed formation and metabolism.

- **immature soil**
a soil that lacks clear individual horizons.

- **immersed**
(L. in, into, mergere, to dip, plunge) growing under water.

- **immortelle**
any one of the flowers called everlasting. These flowers feel papery and keep their natural colour and appearance after drying, therefore they are widely used for wreaths and other floral arrangements.

- **imparipinnate**
having pairs of leaflets opposite each other along a central stem, with a single leaf at the end. Pinnate with a conform terminal leaflet.

- **imperfect**
1. flowers lacking either pistils or stamens.
2. d*iclinous*; having the pistils and stamens in separate flowers. Flowers having only male or female parts.

- **imperfect flower**
a flower containing stamen and pistil organs required for pollination but lacking sepals or petals or both of these organs.

- **impervious soil**
a soil through which water, air or roots penetrate very slowly or not at all.

- **impressed**
sunken (such as veins).

- **impression**
a type of preservation that represents the negative imprint of an organism; compare to compression.

- **impression mark**
a mark on the proximal face of a pollen grain retained from the post-meiotic stage. This mark can be linear from tetragonal tetrads or Y-shaped from tetrahedral tetrads. Examples: Nypa fruticans, Howea belmooreana (Palmae).

- **inaperturate**
describing a pollen grain or spore without apertures. Example: Populus (Salicaceae), Taxus (Taxaceae).

- **inarable**
not arable, not capable of being plowed or tilled.

- **inarch**
to graft by uniting a scion to the stock, without separating the scion from its parent tree.

- **inarching**
this is a method of grafting new plant growth onto a stronger

root system by establishing young plants (usually, one that is in a pot) near an existing tree and at the point where they meet, removing matching areas of bark. The two cut surfaces are then fitted together and bound with soft tying material until they grow together and can be gradually separated with the new branches attached to the older rootstock.

- **inbreed**
  to breed from individuals of the same parentage or closely related.

- **inbreeding**
  deliberate self-pollination of a plant by transferring pollen from one of its flowers to the stigma of the same flower or to another flower of the same plant. Repeating this process for several generations ensures that the line will breed true.

- **incanous**
  whitish or greyish.

- **incidence / incidences**
  a structure formed by the termination of the inner wall layer at the endogerminal which gives rise to a deep V-shaped atrium that points towards the centre of the grain. Example: Pseudotrudopollis.

- **incised**
  1. sharply and usually deeply and irregularly cut, such as the edge of a leaf.
  2. (L. *in*, into, *caedere*, to cut) with sharp angles between the lobes; having deeply cleft margins.
  3. appearing regularly slashed.

- **included**
  1. not sticking out or projecting, as when pistils are included in the corolla.
  2. (L. *in*, in; *claudere*, to shut, close) not projecting beyond an enclosing part.
  3. not exserted.

- **included veinlets**
  veins ending inside areoles.

- **incrassate**
  vt. (L. *in*, in; *crassus*, thick) becoming thick or thicker, especially toward a tip or margin.

- **incubous**
  overlapping, like shingles.

- **incumbent**
  1. resting or leaning upon. Also applied to a pair of cotyledons curving back down the hypocotyl so that the back of one is turned towards the hypocotyl and the back of the other is turned away from it.
  2. lying on another organ, such as an incumbent anther lying against a filament.

- **incurved**

applied to *petals* and *florets* that curve inward to form a compact, rounded shaped.

- **incurvispinus**

with inwardly curved spines.

- **indeciduous**

not deciduous, not losing its leaves annually, evergreen, as trees.

- **indehiscent**

1. failing to split open at maturity.
2. fruits that don't burst open; not popping or splitting open spontaneously when mature, as a capsule or an anther.
3. a (L. *in*, not; *dehiscens*, gaping) fruits which do not open to release seeds, but whole fruit is shed from the plant; not opening to release spores.

- **indeterminant growth**

a type of growth characterised by production of an unlimited number of organs.

- **indeterminate**

a plant part that continues to grow while flowering, such as indeterminate tomatoes, which continue to grow until they die or a flower spike that continues to grow longer while its lower blossoms open.

- **index fossil**

a geographically widespread fossil, that is diagnostic of a particular time period and therefore useful in correlating the age of rock formations from different geographic areas.

- **indigenous**

1. occurring naturally in an area. Not introduced.
2. Native, not exotic.
3. (L. *in*, within; *gignere*, to bear, produce) native originating or occurring naturally in the place specified.

- **indoor gardening**

the art of growing plants indoors using natural and/or artificial light.

- **indumentum**

1. thick hairy covering.
2. the surface hairs of a plant or constituent part, especially where dense.

- **induplicate**

rolled inward, as the leaves in a bud.

- **indusium**

a small flap covering the sori or fruit dots, in ferns, a collection of hairs united so as to form a sort of cup and enclosing the stigma of a flower.

- **inermis**

of a plant without spines (Latin).

- **inferior**
  below another organ or part. Ovary is embedded below the point of attachment of the perianth and anthers.

- **inferior indusium**
  an indusium attached beneath the sorus with the sporangia appearing above it.

- **inferior ovary**
  a basal, ovule-bearing part of a pistil borne below the point of attachment of the perianth and stamens.

- **infiltration rate**
  the rate of water controlled by surface conditions that enters the soil.

- **inflated**
  enlarged or swelled out, as an inflated perianth. Bladdery, swollen and transparent or appearing so.

- **inflorescence**
  1. a flower cluster, including bracts, on a stalk.
  2. a group of flowers borne on a single axis (stem). Examples are: umbel, spike, panicle, corymb and so on.
  3. (L. *inflorescere*, to begin to blossom) a flower or putting forth blossoms, the mode of development and arrangement of flowers on an axis. A flowering branch.
  4. the complete flowering head of a plant, often with multiple flowers.

- **infra-**
  a prefix meaning below or beneath.

- **infrafoliar**
  below the leaves, as infrafoliar flowers.

- **infrared mulch**
  a plastic sheet that is placed around plants to warm the soil by infrared radiation and to inhibit weed growth by blocking most visible radiation.

- **infraspecific**
  referring to categories of classification below that of species, such as subspecies.

- **infratectum / infratecta / infratectate**
  a general term for the layer beneath the tectum, which may be alveolar, granular, columellar or structureless. Synonym of interstitium .

- **infraturma / infraturmae**
  an artificial grouping of form-genera of spores and pollen in the turma- system of Potonié. See **anteturma**, **subturma**, **turma**.

- **infructescence**
  1. the fruiting stage of a plant.

2. (L. *in*, into; *fructus*, fruit) the inflorescence in a fruiting stage, collective fruits.

- **infundibular**
  funnel-shaped.

- **infundibuliform**
  having the form of a funnel.

- **ingraft**
  to insert, as a scion of one tree or plant into another, for propagation. Also spelled engraft.

- **inner tetrad mark**
  a tetrad mark at the inner layer of a spore and which does not reach up to the margin when seen in polar view and always smaller than the outer tetrad mark.

- **inordinate**
  describing a pollen grain or spore with an arrangement of elements irregularly distributed.

- **inorganic**
  1. of molecules or chemistry based mainly on elements other than carbon.
  2. not natural, artificial., composed of other than animal or plant matter or mineral.

- **inorganic fertiliser**
  a chemical fertiliser that quickly releases nitrogen into the soil.

- **inorganic nitrogen**
  the element nitrogen in combination with other mineral elements.

- **insecticidal soap**
  a low-toxicity pesticide made from the salts of fatty acids. It kills soft-bodied insects, like aphids, spider mites, thrips and whiteflies, by disrupting cell membranes. (You can make your own by mixing 2 tablespoons dishwashing detergent in 1 gallon of water.) It may burn foliage in hot weather. It is sometimes combined with other materials, such as citrus oil or sulfur, to increase efficiency. Complete coverage of the plant is necessary for it to work.

- **insecticide**
  an agent that kills insects, specifically, that which kills insect pests of plants and humans.

- **insectivorous**
  trapping and obtaining nutrients from insects. Plants that are specially adapted for catching insects. Most, but not all, are grown for interest rather than beauty. Once their prey is captured, the plants secrete a fluid, which breaks down the insect's bodies so that they may be absorbed as plant food. The following are few

examples of insectivorous plants: Dionaea (Venus's Flytrap), Nepenthes, Sarracenia (Pitcher Plant), Darlingtonia (California Pitcher Plant) and Utricularia (Bladderwort).

- **inserted**
  attached, as a stamen on a corolla.

- **insertion**
  the place or manner of attachment, as the insertion of petals.

- **insula / insulae / insulate**
  synonym of areola.

- **intake rate**
  the rate at which rain or water enters the soil. Expressed in inches per hour.

- **intectate**
  describing pollen grains without a tectum, but with sculpturing. Examples: Viscum (Loranthaceae), Ilex (Araliaceae). See **atectate, etectate, semitectate, tectum**.

- **integument**
  1. the covering of an ovule or other structure.
  2. the outermost layer(s) of an ovule which will develop into the seed coat, most seed plantovules have one integument, angiosperm ovules have two integuments (see **double integument**).
  3. the outer envelope of an ovule, which, with other parts, forms the seed coat.
  4. a natural outer covering as husk or rind.
  5. the outer protective layer of the ovule.

- **inter-**
  between or among. (ME. *enter-*, *inter-*; OFr. *entre-*, *inter-*; L. *inter-* from *inter*, prep. between, among, during) a prefix meaning between, among - as intercellular. A prefix for in between.

- **interaspidium / interaspidia**
  the area of exine between (aspides. Example: Dorstenia (Moraceae).

- **interaxillary**
  situated between or within the axils of leaves.

- **intercellular**
  adj. lying between cells, as intercellular space in plant tissue.

- **interfascicular cambium**
  arises between vascular bundles.

- **interference**
  (L. *inter*, between; *ferire*, to strike) the overall influence of one plant or groups of plants on another and encompasses allelopathy or competition or both of these processes.

- **interfoliaceous**
situated between the leaves, especially opposite leaves. Also interfoliar.

- **intergenicula**
nonflexible, calcified segments between flexible joints in articulated corallines and Halimeda.

- **intergrading**
merging from one form to another by a continuous series of intermediate forms rather than divided into discrete sets.

- **interlacunar gaps**
gaps in the short sexinous ridges which often divide the ectoapertures of lophate pollen grains into lacunae. Example: Cichorium intybus (Compo-sitae).

- **interlacunar ridges**
ridges separating the lacunae in lophate pollen grains. Example: Cichorium intybus (Compo-sitae).

- **Interloculum / interlocula**
a space between the inner- and outer wall layers that is present around most or all of the pollen grain. Example: Extratriporopollenites conjun-ctus.

- **internal foramen / internal foramina**
foramen present within the sexineectexine. See **foramen.**

- **internal tectum**
a more or less continuous layer within the outer sexineectexine composed of laterally connected parts of columellae. Example: Nigella damascena (Ranunculaceae).

- **internodal (n)**
pertaining to sections of stem between leaves or buds.

- **internode**
part of a stem between two nodes.

- **internode**
1. the segment of a plant stem between the points where leaves are attached.
2. the part of stem that lies between two *nodes* or joints on a plant.
3. part of axis intermediate between two nodes.
4. (L. *inter,* between; *nodus,* knot) the portion of a stem between nodes.
5. section of stem between leaves or buds.

- **interporal lacuna / interporal lacunae**
a lacuna occurring between the ends of two ectoapertures

and adjacent to the polar area of a lophate pollen grain. Example: Tragopogon pratensis (Compositae). See **lacuna.**

- **interporium / interporia**
  the meridional segment between two pores extending to the poles. See **apoporium** and **mesoporium.**

- **interradial**
  referring to areas of the proximal face or the equatorial periphery of trilete spores, lying between the radial arms of the laesurae.

- **interspecific competition**
  competition between species for nutrients, space, light etc.

- **interstitium**
  the layer of the exine situated between the nexine and the tectum. Synonym of infratectum.

- **intertextus**
  interwoven.

- **intertidal**
  occurring between high and low tide levels, exposed at low tide.

- **intexine**
  synonym of nexine.

- **intine**
  1. the innermost layer of the wall of pollen and spores made of cellulose and pectates, compare to exine
  2. the innermost of the major layers of the pollen grain wall underlying the exine and bordering the surface of the cytoplasm.

- **intra-**
  within or inside.(L., from *intra*, within, inside) a combining form meaning within, inside of, as intracellular. a prefix meaning within. See also: **infra-.**

- **intracellular**
  being or occurring within a body cell or within the body cells.

- **intrareticulum / intrareticulate**
  synonym of infrareticulum.

- **intraspecific competition**
  a type of competition whereby an individual plant competes with one or more members of the same species for nutrients, space, light etc.

- **intrazonal**
  a term used to describe a soil group with characteristics determined by local factors, such as parent material.

- **introrse**
  turned or facing inward, such as an introrse anther facing toward the center of the flower.

- **involucel**

1. a secondary involucre in a compound cluster of flowers, as in many of the Umbelliferae.
2. (L. *involucrum*, covering) a secondary involucre, as the bracts subtending the secondary umbels in the Umbelliferae.

- **involucral**

pertaining to an involucre.

- **involucrate**

having an involucre.

- **involucre**

1. a whorl of small leaves close underneath a flower or flower cluster.
2. a collection of whorls of small leaves or bracts beneath a flower or an inflorescence.
3. sterile jacket of cells around reproductive structures.
4. a group of closely placed bracts that subtend or enclose an inflorescence.

- **involucriform**

resembling an involucre.

- **involute**

1. rolled inward, as the edge of a leaf.
2. (L. *involutus*, rolled up) leaves having the edges rolled inwards at each side, toward the adaxial side.

- **involution**

(L. *involutus*, rolled up) a rolling inwards of leaves.

- **iridescence**

rainbow-like display of colour on an alga's surface.

- **iron**

plants need this soil nutrient to manufacture chlorophyll. In *alkaline* soil, plants may not be able to absorb iron from the soil, which causes the leaves to turn yellow with green veins. Products, containing iron chelates can be added to the soil or sprayed on the plant's leaves to make iron available to the roots and foliage.

- **iron chelate**

a form of iron that is readily available to plants when iron in the soil isn't available to the roots. Products that contain iron chelates can be added to the soil or sprayed on the foliage.

- **irradiance**

the amount of light energy illuminating the Earth per unit area.

- **irregular**

not having all the members of the same part alike, asymmetrical, lacking uniformity (often refers to flowers.)

- **irregular flower**
  a flower in which similar parts are unequal in size, form or extent of fusion as in e.g. *Leguminosae* (pea family). Often bilaterally symmetrical.

- **irregular symmetry**
  irregular flowers are bilaterally symmetrical, they are divisible into equal halves about 1 plane only.

- **irrigate**
  to pass a liquid over or through, specifically, to water by causing a stream or streams to flow over land.

- **irrigation**
  the act of watering or moistening; the covering of anything with water or other liquid for the purpose of making or keeping it moist, such as the distribution of water over land to promote the growth of plants.

- **isadelphous**
  having the separate bundles of stamens equal in number, as in the flowers of beans.

- **isandrous**
  having the stamens similar and equal in number to the divisions of the corolla.

- **isanthous**
  having regular flowers.

- **isodiametric tetrad**
  a general term for tetrads in which all members are more or less the same in size.

- **isogamous**
  having gametes which are morphologically alike.

- **isogynous**
  having the pistils or the carpels of which the single pistil is composed, equal in number to the sepals.

- **isomerous**
  composed each of an equal number of parts, as the members of the several circles of a flower; exhibiting isomerism.

- **isomorphic**
  morphologically similar, having similar looking gametophyte and sporophyte phases of life history.

- **isopolar**
  describing a pollen grain or spore in which the proximal and distal faces of the exine are alike.

- **isospore**
  a spore of a plant producing only one kind of spore. See **miospore**.

- **isosporous**
  describing plants producing only one kind of spore.

- **isostemonous**
  having the stamens equal in number to the sepals or petals.
- **isostemony**
  the state of being isostemonous.
- **isotropic**
  having the same property in all directions.
- **jardiniere stand**
  an ornamental structure for holding plants.
- **joint**
  a pont where a side-shoot or leaf grows out of a stem. A single section or segment of a stem composed of a series of units.
- **jointed**
  with *nodes*, as the joints of a grass stem. With stems that can be pulled apart easily at the nodes, as in *Equisetum*.
- **jugum**
  a pair of opposite leaflets in a pinnate leaf. A ridge on the fruits of the family Ammiaceae.
- **juvenile**
  young or immature, not having flowered.
- **juvenile leaves**
  the set of leaves, especially when differing from the mature leaves.
- **kainite**
  a natural salt contained in a fertiliser that is used as a source of potash (14% k).
- **kaolin minerals**
  a group of nonswelling clay minerals in which one layer of silicon and oxygen alternates with a layer made up of aluminum, oxygen and hydrogen.
- **karyogamy**
  union of two nuclei.
- **karyokinesis**
  the process of division of a cell nucleus, distinct from cytokinesis division of the whole cell.
- **karyotype**
  the gross morphology of the chromosomes.
- **keel**
  1. a central ridge along the back of any organ of a plant.
  2. the lowest, fused petals of a butterfly-shaped flower.
  3. (AS. *ceol*, ship) the folded edge or ridge of any structure.
- **keiki**
  a vegetative offshoot (plantlet) formed at a node, as on an inflorescence (of an Orchid).
- **kelp**
  member of brown algal order Laminariales.

- **kernel**
  the inner portion of a seed, usually refers to an edible seed or the central part of a nut or fruit seed. The whole seed of a cereal such as corn or wheat. To form kernels.

- **kerogen**
  a solid complex organic material which yields petroleum type hydrocarbons under heat and pressure.

- **kingdom**
  the highest grouping of similar organisms within the system of classification. Terrestrial life is divided into five Kingdoms. Animalia (animals), Plantae (plants), fungi, Protista (protozoans and eucaryotic algae) and Monera (blue-green algae).

- **knee**
  a sharp bend in a tree branch. A protruding enlargement of the roots at the base of certain swamp-growing trees.

- **kneeler**
  a device to help gardeners lower themselves down or push themselves up from a kneeling position. It usually has handholds several feet high attached to a cushioned platform. It may be reversed and used as a garden bench.

- **kneeling pads**
  a usually rectangular and waterproof mat that is used for cushioning the knees while gardening.

- **kneepads**
  cushions that are strapped behind the knees to protect the kneecaps while kneeling or crawling.

- **knock-outs**
  a plant in which a gene has been rendered non-functional. By eliminating the gene's function, the investigator can infer its role in the plant's physiology from the phenotype.

- **knot**
  a lump or swelling in or on a part of a plant, such as the node of grass or the thick part of the wood of a tree where a branch grows out of the trunk.

- **knurl**
  a thick, deformed knot in wood.

- **kyrtome**
  a more or less arcuate fold or band in the interradial areas outside the (laesurae of trilete spores. Examples: Ahrensi-sporites guerickei, Concavi-sporites rugulatus.

- **labellum**
  a lip, particularly the lip of an orchid. The median, modified petal of an Orchid flower.

- **labiate**

1. two-lipped, as a calyx or corolla divided into two unequal parts, like the lips of a mouth.
2. (L. *labium*, lip) lipped, as in a calyx or corolla.
3. of a monopetalous corolla or calyx where divided into unequal divisions.

- **labium**

a lip. The lower lip of a flower with two lips.

- **labrum / labra / labrate**

1. the elevated and/or thickened part of a laesura between the commissure and the remainder of the proximal surface. See **margo.**
2. protruding part of the exine at the pores of porate pollen grains. Example: Lusatipollis.

- **laccate**

appearing as if varnished.

- **lacerate**

1. appearing torn as the edge of a sepal or petal; irregularly cut or cleft.
2. (L. *lacer*, mangled, lacerated) said of a margin torn irregularly.

- **lacerated**

appearing irregularly slashed.

- **lachrymiform**

tear-shaped, like pyriform, but with no contraction towards the point.

- **lacinia**

a long slash or incision in a leaf, petal or similar organ; one of the narrow lobes resulting from such cuts.

- **laciniae (n)**

an irregular narrow strip.

- **laciniate**

1. irregularly cut into narrow-pointed lobes; fringed; refers to petals, bracts, leaves etc. Also laciniated, laciniose.
2. (L. *lacinia*, a hem) cut into narrow, jagged lobes or segments.
3. fringed with irregular strips or segments; very deeply incised.

- **lacinula**

a small, narrow lobe or fringe.

- **lacinulate**

having small, finely cut leaves or petals.

- **lacrimiform**

teardrop-shaped.

- **lacteus**

milky.

- **lacuna**

1. a hole or cavity.
2. chamber or internal air space.
3. a cavity or depression

4. a depressed area surrounded by ridges in lophate pollen grains. See **abporal lacuna, circumpolar lacuna, equatorial lacuna, interporal lacuna, paraporal lacuna, polar lacuna, poral lacuna.**

- **lacunar collenchyma**
cell wall is thickest in the corners, intercellular air spaces present.

- **lacunate**
(L. *lacuna*, cavity) with air spaces or chambers in the midst of tissue.

- **lacunulose**
having tiny holes.

- **laesura / laesurae, suffix -lete**
the arm of a proximal fissura or scar of a spore.

- **laevigate**
a general term for smooth, as if polished. Synonym of psilate.

- **lalongate**
describing the shape of a transversely elongated endoape -rture. Example: Filipendula (Rosaceae). See **lolongate.**

- **lamella**
a thin plate or an organ or part resembling one. Raised, membranous outgrowths on the *labellum*, as in the group, *Coelogyne*. Also lamel.

- **lamellar**
terminated by two small plates.

- **lamellar collenchyma**
cell wall is thickest on two opposite sides.

- **lamellate**
composed of or having lamellae. Also lamellar.

- **lamellate**
1. made up of thin plates or lamina.
2. terminated by two small plates.

- **lamina**
1. the blade or expanded part of a leaf.
2. a thin plate or scale. Specifically, the blade or expanded part of a leaf or petal.
3. the leaf tissue other than the veins or axes.
4. flattened leaf-like blade.
5. (L. *lamina*, a thin peice of metal or wood) the expanded, blade part, of a foliar leaf, petal etc.
6. a thin plate or scale, the surface of a leaf excluding the petiole.

- **laminaran**
storage product of Phaeophyta, branched polymer of polysaccharide made up of glucans, soluble in the cell.

• **laminarin**
a ß1,3 glucan reserve.

• **lanate**
with long, soft wooly hairs. Woolly, covered with long, fine, curly hairs.

• **lance-linear**
long, thin and pointed. (i.e. a lance-linear leaf).

• **lanceolate**
shaped like a long narrow lance with a gradual taper from the base to the tip.

• **lanceolate**
1. long and thin and broadest below the middle, tapering to a point like a lance, lance-shaped.
2. (L. *lancea*, a lance) shaped like a lance, broadest toward the base and narrowed to the apex, several times longer than wide.
3. tapering in shape, like a spear-head, eliptical with a length: breadth ratio of between 3:1 and 6:1.

• **lanciform**
lance-shaped, long, narrow and pointed, sometimes broader at the base than at the tip.

• **land**
refers to the solid surface of the earth. A broad term that refers to soil, surface soil, mineral deposits and water supplies in a certain area.

• **land-capability classification**
soil that is grouped into special units, subclasses and/or classes according to their capability for intense use and treatments that are required for sustained use.

• **landform**
a geomorphological term used to express earthly surface formations.

• **landscape**
a view or prospect of rural scenery, more or less extensive, such as seen from a single point of view.

• **landscape fabric**
any of various porous organic or inorganic cloths or mats used to discourage weeds from growing. Also called weed mat.

• **landscape-gardening**
the art of laying out grounds and arranging trees, shrubs, flowers, paths, fountains etc. to produce an attractive surrounding.

• **lanose**
woolly, lanate.

- **lanuginous**
woolly or downy, covered with soft fine hairs.

- **lanugo**
soft, thick downy growth.

- **larva**
the first stage of an insect after hatching from the egg. (i.e. A caterpillar is a larva of a butterfly or moth.) The larvae are often more serious pests than the adults.

- **lasiacantha**
with shaggy or hairy spines.

- **lateral**
growing or positioned at the sides. A side shoot that grows from a shoot or root, side growths, which develop from the main branches of a tree, shrub or other plant.

- **lateral leaf**
leaf on the side of the stem, as in heterophyllous species of *Selaginella*.

- **lateral lobes**
the two lobes to either side of the midlobe of a three-lobed *labellum*.

- **lateral root primordium**
arises in the pericycle.

- **lath house**
a framework of widely spaced narrow boards used to protect plants from the sun by providing partial shade.

- **laticifer**
1. cells containing a characteristic milky fluid called *latex*.
2. a secretory structure that produces latex.

- **laticiferous**
bearing, containing or secreting latex.

- **latimurate**
describing a pollen grain or spore with broad muri.

- **latiporate**
describing pollen grains with pores in one hemisphere only. Examples: Juglans, Carya (Juglandaceae.)

- **lawn**
space, usually level, covered with grass and kept smoothly mown, such as in front of a home etc.

- **lax**
loosely or widely spaced, as a lax cluster of flowers.

- **layer**
a general term. Applied in palynology to any distinct stratum of the sporoderms.

- **layering**
a method of propagation in which a shoot of a living stem is bent and pegged down to

the soil. The shoot develops roots while still attached to the parent plant; once established, it is cut from the mother plant to form a separate plant, which can then be potted up. The illustration shows a shoot bent down and forming roots below the soil's surface. A stake supports the end of the shoot.

- **leach**

to wash or drain by trickling water through; to treat by downward drainage, as rain washing away nutrients from the soil.

- **leaching**

the process of losing nutrients or salts from the soil through percolation or treating by downward drainage.

- **leader**

the main or terminal shoot of a plant, such as the topmost point of a fir or larch tree or the strongest shoot of a young fruit tree.

- **leaf**

1. an outgrowth from a node on a plant's stem. Leaves come in many shapes and sizes, but are often flattened to maximise their surface area for capture of energy from sunlight (photosynthesis) using chlorophyll contained in chloroplasts.

2. a flattened, photosynthetic structure of a plant arranged on a stem.

3. these grow along the sides of branches or stems or at the bases of the plants. Most are green and contain chlorophyll, though they vary in their shapes and sizes. They are the part of the plant that ordinarily performs photosynthesis, the process that converts sunlight and carbon dioxide into energy. It is common for leaves to grow, fall off and be replaced. Plants entering dormancy may lose all their leaves. Maples, for example, drop their leaves in autumn and replace them in the spring and summer.

4. a lateral appendage of the stem produced by the shoot apical meristem.

5. (ME. *leef* fr. OE. *leaf*, akin to OHG. *loub*, leaf, foliage) a lateral outgrowth from a stem that constitutes part of the foliage of a plant and functions primarily in food manufacture by photosynthesis.

6. usually flattened, usually green plant structure being most often the primary organ of photosynthesis.

- **leaf cutting**

a *cutting* made from a single leaf.

- **leaf dimorphism**
leaves of two distinct types produced by the same plant.

- **leaf gap**
1. a parenchyma filled interruption in a stemís cylinder of vascular tissue immediately above the point at which a branch of vascular tissue (leaf trace) leading to a leaf occurs.
2. in a siphonostele, a break in the vascular cylinder above the point where a leaf trace arises.

- **leaf mould**
this is a compost of decayed leaves. It is made by piling leaves of deciduous trees and shrubs together and allowing them to rot. The pile should be trodden firm and the leaves should be wet. Every 6 months or so, the pile should be turned over so the outer section of the pile is in the centre. It takes a year and a half to two years for the leaves to decompose enough to use for leaf mould.

- **leaf plant**
a plant grown for its decorative foliage.

- **leaf primordium**
arises at the shoot apical meristem.

- **leaf scar**
a mark left on the stem where a leaf has been lost or removed. It can easily be seen on Palms where old leaves have been shed.

- **leaf scorch**
this may be caused by dryness, high temperatures, disease or lack of certain elements in the soil. The foliage looks as if its been dried by heat.

- **leaf trace**
1. a branch of vascular tissue leading from the main vascular cylinder of the stem to a leaf.
2. a vascular bundle that connects a leaf to the axial vascular system.

- **leafbud**
a bud producing a stem and leaf, unlike a flowerbud, which contains a blossom.

- **leaflet**
part of a compound leaf.

- **leafstalk**
the stalk that attaches a leaf to the stem or branch. The *petiole*.

- **lectin**
1. a diverse group of plant proteins capable of aglutinating cells (such as red blood cells). Lectins contain sites.
2. that recognise and bind to sugar residues, such as those in-

cluded in the structure of glycoproteins. Ricin is a notorious toxic lectin from the seeds of the castor oil plant.

- **lectotype**
  a specimen selected from the original type material to serve as the nomenclatural type when a holotype was not originally designated or is missing.

- **leggy**
  of plants exhibiting excessive, rank growth, often caused by excessive watering, application of fertiliser or insufficient light.

- **legume**
  1. a fruit with a single compartment that splits along two sutures or seams e.g. pea pod. A plant in the family Leguminosae (peas, beans) that can fix nitrogen from the soil due to the presence of symbiotic bacteria (e.g. *Rhizobium species*,) in root nodules.
  2. (L. *legere*, to gather) a 1-locular fruit, usually dehiscent along two sutures, bearing seeds along the ventral suture. A leguminous plant.

- **lemma**
  (Gr. *lemma*, husk) the lower (abaxial) and larger, of two membranous bracts enclosing the flower in grass.

- **lentic**
  still bodies of water.

- **lenticel**
  1. a pore in tree bark through which gas exchange occurs.
  2. a lens-shaped body of cells, formed on the outside of a woody plant stem, which serves in the exchange of gases between the stem and the outer air.
  3. a region of the periderm where cells are loosely packed, allows gas exchange.
  4. (L. *lens, lentis*, lentil) corky spots on young bark, arising in relation to epidermal stomates.

- **lenticellate**
  pertaining to or having lenticels.

- **lenticular**
  1. shaped like a biconvex lens.
  2. resembling a lentil or a double-convex lens, as some seeds.
  3. (L. *lenticula*, a lentil) shaped like a double-convex lens.
  4. lens-shaped, resembling a double convex lens.

- **lentiform**
  lens-shaped, resembling a double convex lens.

- **lentiginous**
  covered with minute dots, freckled, speckled.

- **lepidote**
  covered with dry scales or scaly spots.

- **leprose**
  having scaly or dry scales.

- **-lept**
  a suffix for thin, indicating that a pollen grain has a leptoma.

- **leptocaulis**
  with slender stems.

- **leptoid**
  photosynthate conducting cells of bryophytes, compare to hydroid.

- **leptoma**
  a thin area at the distal pole of a pollen grain, presumed to function as an aperture. Example: Pinus (Pinaceae).

- **leptosporangiate**
  having the entire sporangium develop from a periclinal division of a superficial cell or small group of cells.

- **leptosporangium, leptosporangiate**
  leptosporangium: sporangium developed from a single superficial cell. leptosporangiate: having leptosporangia. This is a derived character state in the fern clade. Compare to eusporangium

- **-lete**
  a suffix to denote the presence (or absence) of laesura. Examples: alete, monolete, trilete.

- **leucoplast**
  colourless, usually starch-containing plastid.

- **level**
  to prepare soil for planting by raking gently.

- **levigate**
  orthographical variant of laevigate.

- **liana**
  liane- a woody, climbing vine. a climbing and twining, usually woody, plant found mostly in tropical rainforests, which may climb to the level of the tree canopy. Also liane.

- **libriform fibre**
  a cell in the xylem that is very long and thin and has simple pits, sometimes called 'xylary fibres' to distinguish them from extraxylary fibers, which look similar, but have a different evolutionary origin.

- **lichen**
  a complex differentiated organism formed by a symbiotic relationship between an algae and a fungus.

- **lid**
  the cap of a box-like seed capsule.

- **light soil**
  a term used to indicate a sandy or coarse-textured soil that drains rapidly. It also may lack important plant nutrients.

- **ligneous**
  consisting of or resembling wood, woody, as distinguished from herbaceous.

- **lignescent**
  tending to be or become ligneous or woody, somewhat woody.

- **lignify**
  1. to become wood or woody.
  2. vt., to convert into wood or woody tissue; to become wood or woody by chemical and physical changes in the cell walls that convert some or all of the constituents into lignin or lignocellulose.

- **lignin**
  1. a diverse group of non-toxic organic polymers often associated with cellulose in the walls of many plant cells, especially in woody plants. In this form, lignin acts as a cement to exploit the strength of cellulose fibres, while allowing flexibility. The mechanical properties of this natural composite material (wood) compare favourably with many man-made composite materials such as fibre-glass. After cellulose, lignin may be the second most abundant biopolymer in the world with an estimated annual production of 10^8 tons, but see also chitin. Lignin is formed by the irreversible removal of water from sugar monomers to create phenolic aromatic structures. Polymerisation and extensive cross-linking occur to create very large three-dimensional lignin molecules. Sometimes lignin is isolated as a brown powder, but more often as a gummy mixture of lignins with a wide range of molecular weights. Lignin resists attack by most microorganisms and anaerobic processes tend not to attack the aromatic rings at all. Aerobic breakdown of lignin is slow and may take many days. Hence, the presence of lignin strongly affects the biodegradeability of woody materials.
  2. an aromatic polymer that rigidifies may secondary cell walls. Ligin is stained red by phloroglucinol solutions.

3. (L. *lignum*, wood) organic substances which act as binders for the cellulose fibers in wood and certain plants and adds strength and stiffness to the cell walls. Chemical structure of lignin is composed of a polymer of high carbon content but distinct from the carbonates. Consists of C6, C3 units.

- **ligniperdous**

destructive of wood.

- **lignum**

woody tissue, as contrasted with soft tissue or with bark.

- **ligulate**

1. strap-shaped, such as a corolla, leaf or petal. Furnished with ligule, as a ligulate grass.
2. (L. *ligula*, little tongue) having or pertaining to ligules.
3. having strap-like florets.

- **ligulea**

1. a tiny, tongue-like appendage on the adaxial surface of leaves of some members of the lycopods.
2. a strap-shaped plant part; the blade formed by the corolla in some or all the florets of many members of the daisy family, Compositae. The strap-shaped part that sticks up from the top of the sheath in many grasses, palms and some other plants.
3. a small membranous outgrowth or projection at the base of the leaf, appearing above the sporangium in fertile leaves, as in *Selaginella* and *Isoetes*.
4. hyaline extension of the leaf sheath on the adaxial side of the leaf.

- **liguliflorous**

having heads composed exclusively of strap-shaped florets.

- **limb**

1. a primary branch of a tree.
2. the expanded, flat portion of a plant part, such as the upper portion of a petal.
3. (AS. *lim*, limb) the spreading part of a synsepalous calyx or sympetalous corolla, usually referring only to the calyx or corolla lobes, sometimes to their lips.
4. synonym of equatorial outline.

- **limbate**

bordered (i.e. a flower in which one colour is edged with another colour).

- **limbus**

a sharp narrow crease of the saccus or pseudosaccus where the outer and inner exine layers are fused. Example, Nuskoisporites dulhuntyi.

- **lime requirement**
the amount of standard ground limestone required to bring a layer of an acre of acid soil (6.6 pH) to some specific lesser degree of acidity. Expressed in tons per acre of nearly pure limestone.

- **lime sulphur**
a caustic pesticide made from calcium polysulfide. Usually used as a dormant spray to control various diseases, such as leaf spot, peach leaf curl and powdery mildew. Also effective in controlling mites, scale and some other insects.

- **limnology**
the scientific study of physical, chemical, meteorological and biological conditions in fresh waters.

- **linear**
1. long and narrow with parallel or nearly parallel sides.
2. long and very narrow with parallel sides, such as a linear leaf. Can also be used in combination, as a linear-lanceolate leaf, which is a leaf that is long, narrow and pointed.
3. (L. *linea*, line) long and slender with parallel or nearly parallel sides.
4. narrow with untapering margins and a length:breadth ratio between 6:1 and 3:1.

- **linear tetrad**
a uniplanar tetrad in which the four members are arranged in a row. Example, Typha spp. (Typhaceae).

- **lineolate**
marked with fine lines. Also lineolated.

- **lingulate**
strap-shaped, longer than linear but shorter than fasciarate, with a length:breadth ratio of between 6:1 and 12:1.

- **lingy**
abounding in heaths, heathy; heathery.

- **linnaeus - carolus linnaeus (1707-1778)**
a Swedish scientist who devised the binomial nomenclature used to identify every living species. The Linnaean System of classification is based on a simple hierarchical structure in which all known organisms are listed according to Kingdom, Phylum, Class order, Family, genus and Species.

- **lip**
1. one of the parts of a bilabiate (two-lipped) flower. In orchids,

it is one of the three petals that has a different shape than the other two. See **Labellum.**
2. (AS. *lippa, lippe,* lip) the upper or lower part of a bilabiate calyx or corolla.

- **lip cells**
the line of cells between which the sporangium dehisces.

- **lira / lirae**
a narrow ridge which forms the murus in a striate pattern.

- **lithocysts**
literally translated 'rock cells', cells containing a granule of calcium carbonate called a cystolith.

- **lithophyte**
growing on rocks. A plant that grows on the surface of rocks, getting its nourishment from the air, as some orchids.

- **lithosol**
a shallow soil consisting of imperfectly weathered rock fragments or nearly all barren rock.

- **littoral**
growing on the beach. Pertaining to a shore, especially of the sea; growing on or near the shore. Near the shore, intertidal.

- **lituate**
forked, with the points turned outward.

- **liturate**
spotted.

- **loam**
the textural class named for soil containing equal amounts of sand, silt and clay, which is usually well-structured, fertile and moisture retentive, but also free draining. A term referring to a mellow soil rich in organic matter.

- **loamy soil**
a general term for a soil of intermediate texture.

- **lo-analysis**
a method for analysing patterns of sexine organisation by means of light microscopy.

- **lobate**
1. having a lobe or lobes, such as a lobate leaf. Also lobated.
2. describing an equatorially aperturate pollen grain with a lobed shape in polar view.

- **lobe**
partial rounded portion of a leaf or other organ, separated from the whole by a deep indentation that does not break the continuity of the structure.

- **lobed**
divided into (usually rounded,) segments.

- **lobed flower**
  a tubular or funnel-shaped flower opening into petal-like segments.

- **lobed margin (leaves)**
  margin indented one quarter or more of the distance from the margin to the midvein or (where this is lacking) to the long axis of the leaf, compare entire margin and toothed margin leaves.

- **lobelet**
  a little lobe, a lobule.

- **lobulate**
  (Gr. *lobos*, lobe) divided into small lobes.

- **locular**
  (L. *loculus*, a cell, box) having the nature of or consisting of cells.

- **locule**
  1. a cavity containing ovules, in a compound ovary there is one locule per carpel.
  2. (L. *loculus*, a cell, box) a compartment of an anther or an ovary.

- **loculicidal**
  pertaining to small separate cavities.

- **lodicule**
  1. one of the two or three delicate scales that occur in the flowers of some grasses, at the base of the ovary and below the stamens.
  2. (L. *lodicula*, coverlet) a scale at base of an ovary in grasses, supposed to represent part of a perianth.

- **loess**
  a geological deposit of fine material, mostly silt, transported by wind.

- **lolongate**
  describing the shape of a longitudinally elongated endoaperture. Example: Rumex spp. (Polygonaceae). See **lalongate.**

- **loment**
  1. a leguminous fruit, which breaks up into many dry, one-seeded pieces when ripe. Also lomentum.
  2. (L. *lomentum*, bean meal) a fruit of some legumes, contracted between the seeds, the 1-seeded segments separating at fruit maturity.

- **long shoot**
  shoots that are distinguished by their widely separated nodes

and and internodes; compare with short shoot.

- **long-day plant**

any plant that needs many hours of sunlight to bloom; usually, plants that are natives to latitudes greater than 60° both north and south of the equator. Long-day plants flower in late spring and early summer. Lettuce is an example, flowering when each day exceeds 17 hours of sunlight. Growers provide long-day plants with long hours of artificial light in greenhouses to force them into out-of-season bloom. A plant that requires more than 12 hours of daylight before flowering will occur.

- **longiaxe**

pollen grains with a polar axis longer than their equatorial diameter.

- **longitudinal**

a general descriptive term, in palynology applied to features which run in lines between the poles.

- **lophate**

describing a pollen grain in which the outer exine is raised in a pattern of ridges (lophae) surrounding depressions (lacunae). Example: Hieracium (Compositae). See **echinolophate, psilolophate.**

- **lopping**

the procedure by which all the branches of a tree, except the leading shoot, are cut off for the sake of the profit to be derived from them. This is in contrast to pruning, in which only some of the branches are cut off for the sake of the tree. That which is cut off; severed branches.

- **lorate (adj)**

lingulate, longer than linear but shorter than fasciarate, with a length:breadth ratio of between 6:1 and 12:1.

- **low maintenance plants**

plants that need little care.

- **lower pitcher (n)**

the form of Nepenthes pitcher found on the lower portion of the plant. See also **upper pitcher.**

- **loxocolp(or)ate**

describing zonocolp(or)ate pollen with ectocolpi arranged so that they converge in pairs.

- **lumen**

intercellular space.

- **lumen / lumina**

the space enclosed by the muri. See **brochus, reticulum.**

- **lumper - coloquial**
  one who takes a broad view of a genus, combining variant forms into individual species.
- **lunate**
  crescent-shaped.
- **lunulate (adj)**
  crescent moon-shaped.
- **lutescens**
  yellowish.
- **lyrate**
  resembling a lyre, for example, a lyrate leaf is one that is divided into several lobes, which increase in size toward one large terminal lobe.
- **lyrate (adj)**
  lyre-shaped, like pandurate, but pinched several times to produce several sets of smaller lobes, e.g: an oak leaf.
- **macroalgae**
  large, macroscopic algae, seaweeds.
- **macrocentra**
  with a large central spine.
- **macromeris**
  with long or large divisions.
- **macrophyllous**
  having large leaves, usually with many veins.
- **macrophyte**
  (Gr. *makros*, large; *phyton*, plant) a member of the macroscopic plant life especially of a body of water; large aquatic plant; the term 'aquatic macrophyte' has no taxonomic significance.
- **macropodous**
  having a long stalk or stem, (of a plant embryo) having a large or long hypocotyl.
- **macropore**
  a large, non-capillary pore found in soil through which water drains by gravity.
- **macrorhiza**
  with long or large roots.
- **macrosclereids**
  column shaped, longer than wide.
- **macroscopic**
  (Gr. *makros*, large; *skopein*, to view) items large enough to be observed by the naked eye.
- **macrospore**
  general term for the larger spores of heterosporous vascular plants (Jackson, 1928).
- **macrothallus**
  large, conspicuous phase of organisms's life history.
- **macula**
  a spot or blotch. Hence Adj. *maculate* = with spots or blotches.

- **macular**
  spotted.
- **maculate**
  a general descriptive term. In palynology it can be applied to pollen grains or spores with a spotted exine, often due to variations in internal exine structure. Example: Maculatisporites.
- **maculation**
  spotting or marking.
- **magnesium**
  an element found in combination in minerals. In plants, located either in the seeds or in the chlorophyll.
- **malathion**
  a man-made insecticide that is effective in destroying a wide range of garden pests, such as aphids, scale and mites.
- **male**
  having flowers that do not produce fruit or seed and only contain stamens.
- **mamilla**
  nipple-shaped tubercle with an areole at its tip.
- **mamillate**
  with breast or nipple shaped projections.
- **mammillate**
  having nipple-shaped protuberances.
- **manoxylic wood**
  wood type that contains abundant parenchyma, typical of cycads, compare with pycnoxylic.
- **marcescent**
  1. withering, but not falling off.
  2. the leaves of short duration, dying at the end of the growing season.
  3. (L. *marcescere*, to wither) withering but remaining persistent.
  4. withering but remaining attached to the plant.
- **margin**
  the edge of a leaf or other organ, edge, rim.
- **marginal**
  along the margin or edge.
- **marginal farmland**
  surface soil that is only a few inches deep. Land repeatedly farmed without benefit of humus or chemical replacements.
- **margo/margine/ marginate**
  an area of exine around an ectocolpus that is differentiated from the remainder of the sexine, either in ornamentation or by difference in thickness. See **annulus, labrum.**

• **marine**
of the sea or saline water.

• **marl**
an earthy deposit consisting mainly of calcium carbonate, used for liming acid soil.

• **marrow**
the pith of plants. The pulp of fruits.

• **marsh**
1. a swampy waterlogged area, either fresh or saline.
2. (ME. *mersh*, meadowland) a tract of wet land principally inhabitated by emergent herbaceous vegetation.

• **massa / massae**
a specialised structure composed of aborted spores and tapetal material found on certain megaspores and often referred to as a floating apparatus. Applied only to megaspores. Examples: Azolla (Azollaceae), Cytosporites varius.

• **massula**
a clump of microspores, as in *Azolla*.

• **massula / massulae/ massulate**
a general term for aggregations of pollen grains dispersed as a unit. Examples: Mimosaceae, Periplocaceae. See **dispersal unit, polyad, pollinium.**

• **mastigoneme**
fine hairlike appendage of a flagellum.

• **matrix**
a medium in which something is embedded.

• **mature**
applied to cells and tissues that have differentiated into their final adult form assuming the function or form characteristic of cells or tissues found in a fully developed part of the plant.

• **meadow**
a piece of low-lying, moist land that is often used for growing hay or as pastureland. An upland area covered with grass and usually surrounded by woods.

• **meal**
consists of canola seed remains after oil has been extracted from them.

• **mechanical analysis**
the physical analysis of soil material to determine its grain-size fraction.

• **median**
in the middle or axial plane.

• **median leaf**
leaf on top of stem, as in heterophyllous species of Selaginella.

- **medicinal plants**
plants that are or have been used medicinally.

- **medine**
a term applied to a faintly lamellated, acetolysis resistant layer considered to be situated between intine and exine. See **mesine**.

- **medulla**
soft internal tissues such as pith. Central tissue of thallus.

- **megagametophyte**
in heterosporous plants and in seed plants. The female gametophyte produced by a megaspore.

- **megaphyll**
a leaf with more than one vein and a leaf trace associated with a leaf gap in the stele; compare to microphyll.

- **megasporangium**
a sporangium that produces megaspores. The sporangium in which megaspores are produced.

- **megaspore**
1. a large, haploid (N=1) spore of a heterosporous plant that produes a megagmetophtye (female gametophyte.)
2. the larger spore produced by plants that bear two types.
3. a spore that gives rise to a female gametophyte.
4. a general term for large spores of heterosporous vascular plants. See **microspore, macrospore.**

- **megasporophyll**
a leaf bearing one or more megasporangia.

- **meiacantha**
with smaller or fewer spines.

- **meiosis**
1. a special type of cell division at the sexual stage of a life-cycle when the diploid or other normal chromosome number is halved in the production of haploid gametes.
2. this is a process occurring during pollination, in flowering plants, in which the chromosomes of the male and female gametes line up together and then divide so as to form new gametes having half the number of chromosomes as the somatic cells of the plant. This process assures that, as traits from both parents are passed on to succeeding generations, the offspring will exhibit some individual variation from either parent.
3. nuclear division in which chromosome number is reduced from diploid to haploid,

genetic segregation of chromosomes.
4. the reduction division of chromosomes.

- **meiospore**
general term for a spore produced by meiosis.

- **meiotic**
pertaining to meiosis: a two stage type of cell division in sexually reproducing organisms that produces gametes with half the chromosome number of the original cell.

- **melanacanthus**
with black spines.

- **melissopalynology**
the study of pollen grains collected by bees and/or found in honey.

- **mellisopalynology**
the study of pollen in honey or other bee products such as 'bee baskets'.

- **meloniform (adj)**
melon-shaped; irregularly sphe -srical with projecting ribs.

- **membrana colpi / membranae colpi**
synonym of colpus membrane.

- **membrana pori / membranae pori**
synonym of pore membrane.

- **membranaceous**
thin, soft and translucent.

- **membrane**
a thin pliable sheet of material that separates and forms the interface between two environments. A thin flexible sheet of tissue connecting, covering, lining or separating various parts or organs. A thin (10 nm) structured bi-layer of phospholipid and protein molecules surrounding every biological cell and intracellular organelles such as chloroplasts, mitochondria and the cell nucleus. Functionally distinct proteins floating in the membranes of cells and organelles control the passage of substances into and out of the enclosed space, catalyse chemical reactions at the membrane surface and allow cell-cell recognition.

- **membranous**
(L. *membrana*, mem- brane) having a thin, soft, pliable texture.

- **meniscoid**
thin, concavo-convex and hemispherical, resembling a watch-glass.

- **mentum**
a 'chin' on some Orchid flowers, formed by the fusion of the lateral sepals and the column-foot.

- **mericarp**
one of the halves of the fruit of certain plants, such as the

maple, whose fruits split into halves.

- **meridional**
describing longitudinal features on the surface of a pollen grain or spore which run along lines perpendicular to the equator.

- **meridionosulcus / meridionosulcate**
a meridional sulcus. Example: Degeneria vitiensis (Degen -eriaceae). See **zonasulcus.**

- **meristem**
1. the growing tissue comprised of stem cells capable of repeated division and differentiation into permanent structures.
2. region of totipotent cells in which cell division and initiation of tissues and organs takes place. See **apical meristem, vascular cambium and cork cambium**
3. in plants, undifferentiated tissue which retains the ability to provide new cells.
4. a defined region where new cells arise in predictable pattern; localised regions of cell division.

- **meristematic region**
a general zone in which cell division is frequent.

- **meristoderm**
superficial, outer layer of dividing cells in brown algae in the order Laminariales.

- **meroplankton**
organisms that spend part of their life cycle as plankton and part on the benthos.

- **mesarch**
a type of xylem maturation in which the protoxylem is embedded in the metaxylem and development proceeds both centripetally (from the outside in) and centrifugally (from the inside out.) Compare to endarch and exarch.

- **mescal**
a strong alcoholic liquor, similar to Tequilla, made by distillation of a fermented preparation of the starchy core of *Agave angustifolia* and other species. Also a common Mexican name for certain *Agaves.*

- **mescaline**
a halucinogenic alkaloid found in *Lophophora williamsii* and other species of cactus traditionally used in native American Indian religous ceremonies and medicine.

- **mesic**
(Gr. *mesos*, middle) conditioned by temperate moist climate; neither xerix nor hydric; per-

taining to conditions of medium moisture supply.

- **mesine**
a laminated, electron-dense layer considered to lie between the intine and the exine.

- **meso-**
a prefix meaning middle.

- **mesoaperture / mesoaperturate**
the middle part of a compound aperture in which there is also an ectoaperture and an endoaperture. Example: Polygonum aviculare (Polygonaceae).

- **mesocarp**
the central portion of the ripened wall of a fruit.

- **mesochile**
the middle part of the lip in certain orchids that have this organ divided into three parts.

- **mesocolpium (pl. mesocolpia) and mesoporium (pl. mesoporia)**
the area of a pollen grain surface delimited by lines between the apices of adjacent colpi or the margins of adjacent pores.

- **mesokaryotic**
having chromosomes that persist in condensed form at all times, the condition of the dinoflagellate nucleus.

- **mesophyll**
1. the photosynthetic layer of tissue of a green plant underneath the epidermis.
2. parenchyma tissue between the upper and lower epidermis of a leaf.

- **mesophyte**
a plant with neither hydromorphic nor xeromorphic features.

- **metabolic pathway**
a series of chemical reactions catalysed by enzymes in a living system. The products of the pathways have vital functions. For example, the Kennedy metabolic pathways found in plants are involved in producing various types of seed oil.

- **metareticulum / microreticulate**
a reticulum wich is characterised by the consistent presence of one porate aperture in each lumen. Examples, Froelichia floridana (Amaranthaceae), Viviania rosea (Vivianiaceae), Kallstroemia maxima (Zygophyllaceae).

- **metaxylem**
a type of primary xylem that differentiates and matures later than the protoxylem, generally metaxylem tracheids are longer than protoxylem.

- **micro-**
  a prefix for small.
- **microcarpa**
  with small fruits.
- **microclimate**
  a small, local climate within a larger climate area, such as a greenhouse or a protected area of a garden.
- **microfossil**
  a fossil or fossil fragment that can only be seen with a microscope.
- **microgametophyte**
  in heterosporous plants and in seed plants: the male gametophyte produced by a microspore
- **micromeris**
  with small parts or divisions.
- **micro-nutrients**
  very small amounts of nutrients needed by plants.
- **microphyll**
  a leaf vascularised by a single vascular bundle that is not associated with a leaf gap in the stele. This type of leaf is typical of the lycopods. Compare to megaphyll.
- **microphyllidious**
  small, leaf-shaped.
- **microphyllous**
  1. having small leaves.
  2. having leaves with only one unbranched vein.
- **micropyle**
  1. a small pore in a seed that allows water absorption. A minute opening in the wall of an ovule through which the pollen tube enters.
  2. a small opening in the integument at the apex of a seed through which either pollen (gymnosperms) or the pollentube (angiosperms) enters.
  3. an opening in the integuments through which a pollen tube enters.
- **microreticulum/microreticulate**
  a reticulate ornamentation consisting of muri and lumina smaller than 1μm.
- **microsperma**
  with small seeds.
- **microsporangium**
  1. produces microspores.
  2. in plants that produce two types of spores, the receptacle that contains only the smaller type, comparable to the pollen sac of the anther in a seed plant.
  3. the sporangium in which microspores are produced.
- **microspore**
  1. a general term for the smaller spores of heterosporous plants, that is, the spores from which

the microgametophyte develops.
2. immature pollen grains that can be effectively triggered and cultured under sterile conditions to produce embryos which can be regenerated into plants (a process called androgenesis.)

- **microsporocyte**
the mothercell of a microspore or pollen grain. See **pollen mother cell**.

- **microsporophyll**
a leaf bearing one or more microsporangia.

- **microstylous**
a flower that has short styles.

- **microthallus**
the tiny, inconspicuous stage of an organism's life history, alternating with the macrothallus stage.

- **midden, pack-rat midden**
midden, garbage or refuse heap. packrats (*Neotoma* sp.) collect and store plant and animal parts in middens, plant material is preserved and cemented together by urin and feces from the packrats, packrat middens are only found from the later part of the Pleistocene through the present, but their fossils have been instrumental in reconstructing Pleistocene and Holocene vegetational changes in the North American southwest and have also been important in archeological studies.

- **midlobe**
the middle lobe of a three-lobed *labellum*, usually brightly coloured and furnished with *calli (callus)*, *keels* and/or *lamelle*.

- **midrib**
the main or central rib of a leaf.

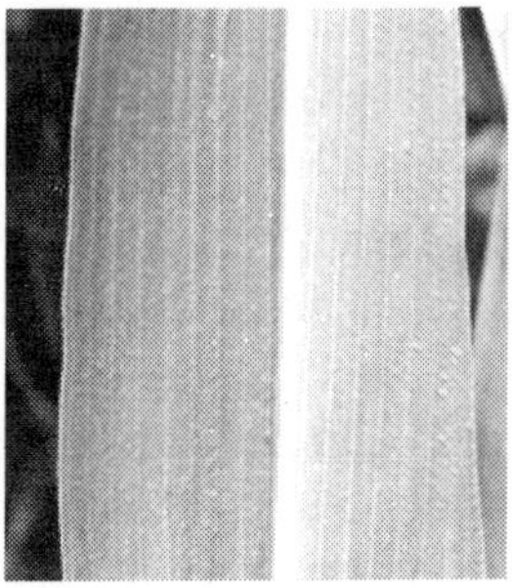

- **mildew**
a growth (usually whitish coloured) produced on living plants by fungus. A fungus that causes such growth.

- **mineral soil**
a general term for soil composed mainly of mineral matter.

- **mineralisation**
the release of mineral matter from organic matter.

- **miospore**
a general term for all fossil plant spores smaller than 200μm, re-

gardless of whether they are isospores, microspores, small megaspores, prepollen or pollen grains. See **microspore, pollen.**

- **mire**

synonymous with any peat-accumulating wetland.

- **mitochondria**

a functional organelle in the cell cytoplasm, of both plants and animals, containing highly ordered membrane-bound arrays of enzymes and proteins involved in respiration and energy metabolism. Mitochondria contain a small circular piece of DNA, which carries genes for a small number of mitochondrial proteins. Mitochondria are inherited almost exclusively from the female parent and DNA sequences encoding mitochondrial genes are analysed to determine interspecific relationships and the slow rates of mutation in mitochondrial DNA sequences are used as one form of 'molecular clock'.

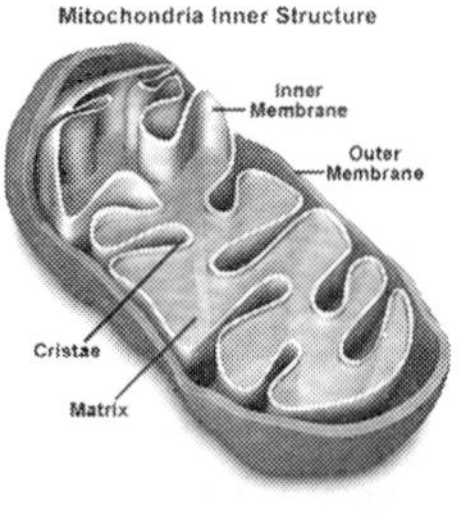

- **mitosis**

1. the normal process of cell division when the diploid or other normal chromosome number is conserved to produce two daughter cells with identical numbers of chromosomes to the parent cell.
2. this is cell division, such as what occurs after fertilisation of the female germ cell by the male. It is usually preceded by halving and differentiation of the chromosomes within the nucleus. Cell division is the process by which plants grow.
3. a general term for nuclear division involving no reduction of chromosomes.

- **mitriform**

having a hollow conical shape, resembling a bishop's miter.

- **modioliform**

hollow, round and depressed, with a very narrow orifice.

- **moisture tension**

the force at which water is held by soil. Expressed as the equivalent of a unit column of water in centimeters.

- **mold**

a three-dimensional preservation type that represents a negative image of the plant; compare to cast.

- **molecular biology**
  the study of the molecules in living organisms, the reactions they undergo, how these reactions are controlled and how molecules are assembled into larger structures such as membranes and fibers. Molecular biology is the study of biology at the level of molecules rather than tissues or cells.

- **molecular farming**
  the application of biotechnology for the selected production of pharmaceutical compounds or other health or industrial compounds within a living organism (e.g. microbe, agricultural crop, livestock).

- **molecular fossil**
  non-structural compounds, for example break-down products of pigments or lignins.

- **molecule**
  a group of atoms that are bonded together.

- **molendinaceous**
  in the form of a set of sails, with multiple projecting wings.

- **molybdenum**
  a trace element present in most soils and plants.

- **monad**
  a pollen grain or spore dispersed as an individual unit, rather than in association with others, such as in a dyad, tetrad or polyad.

- **monadelphous**
  having the stamens of a flower, united into one group, as can be seen in Hibiscus.

- **monander**
  a plant that has only one stamen, such as certain orchids.

- **monandrous**
  having a single stamen.

- **moneocious**
  (one house) producing male and female gametes on the same individual.

- **moniliform**
  1. constricted at intervals and resembling a string of beads, as certain pods, roots and stems.
  2. (L. *monile*, necklace; *forma*, shape) constricted laterally and appearing beadlike.
  3. necklace-shaped; terete with regular bulges.

- **mono-**
  a prefix for one.

- **monoaperturate**
  describing a pollen grain or spore with a single aperture. Examples, monocolpate, monoporate, monosulcat.

- **monocarp**

a plant that bears fruit once and then dies.

- **monocarpellary**

a flower consisting of only one carpel.

- **monocarpic**

1. a plant that flowers only once in its life and then dies.
2. a plant that produces reproductive propagules only once in its lifetime.
3. bearing fruit once and then dying, as annual plants.

- **monocarpous**

a flower that has a single ovary, whether simple or compound.

- **monoclinous**

1. a flower that has both the stamens and pistils.
2. (Gr. *monos*, single, alone; *kline*, bed) having both stamens and pistils in the same flower.

- **monocotyledon**

1. a large group of flowering plants with a single cotyledon that initially emerges from the seed. The *Agavaceae* and *Liliaceae* are monocotyledons.
2. a plant that has a single seed leaf.
3. possessing only one cotyledon.
4. (Gr. *monos*, single; *kotyledon*, cup-shaped hollow) a class of angiosperms having an embryo with only one cotyledon, part of the flower usually in threes, leaves with parallel veins and scattered vascular bundles.

- **monoculture**

growth of a single species in a particular area.

- **monoecious**

1. with both male and female reproductive organs on the same plant.
2. having female and male reproductive structures on the same plant. The reproductive structures can be either unisexual or bisexual. Compare to dioecious
3. having male and female flowers on the same plant.
4. used to describe a species in which a single plant carries both male and female sex organs.
5. (Gr. *monos*, single; *oikos*, house) a plant having unisexual male and female flowers on the same individual, said of a plant having unisexual flowers.
6. having unisexual male and female organs on the same plant.

- **monogynoecial**

formed from a single pistil of a flower, as in certain fruits.

- **monolete**
describing a spore with a single laesura. Example, Dryopteris (Dryopteridaceae).

- **monomerous**
having only one member in each whorl (pistil, stamen, sepal or petal), said of a flower. Monocarpellary.

- **monomorphic**
with a single shape or form.

- **monopetalous**
1. having only one petal. Having the petals united into one piece by their edges or at the base.
2. having a corolla fused into one piece or the petals into a tube.

- **monophyletic**
of a group of organisms sharing and including a common ancestor and including all the descendents of that common ancestor, therefore with similar characteristsics. Otherwise known as a clade.

- **monophyly, monophyletic group**
terms applied to a group of organisms that include an ancestor and all of its descendants. compare to paraphyly, para-phyletic groups and polyphyly and polyphyletic groups.

- **monopodia**
an axis of growth that extends in the direction of previous growth, with lateral branches being produced beneath it, for example, as in certain coniferous trees.

- **monopodial**
1. having the growth of the stem or rhizome continuing in the same direction indefinitely. A habit of growth in which new leaves form from the same growing point.
2. having one main axis of growth.
3. mode of development where primary axis maintains main line of growth and secondary laterals are produced off that main axis.

- **monosaccate**
1. a pollen grain with one saccusor buoyant bladder.
2. describing a pollen grain with a single saccus. Example, Florinites antiquus.

- **monosepalous**
having the sepals united. Having only one sepal.

- **monosperm**
a plant that produces only one seed.

- **monospermous**
  having only one seed.

- **monosporous**
  having only one spore.

- **monostelic**
  having one stele, compare to polystelic.

- **monostichous**
  1. arranged in a single row along a stem, as certain flowers.
  2. arranged in one row or layer.

- **monostromatic**
  blade composed of a single layer of cells.

- **monostylous**
  having only one style.

- **monosymmetrical**
  capable of being bisected into similar halves with reference to a single plane. Bilaterally symmetrical, as in flower of the sweet pea.

- **monotocous**
  producing fruit only once, as annuals.

- **monotypic**
  1. a genus with a single representative species.
  2. a group consisting of a single species.
  3. (Gr. *monos*, only; *typos*, type) a plant of only one type.
  4. represented by a single example.

- **monovalent cations**
  ions having a positive charge.

- **monstrose**
  1. monstrous, an abnormal uncoordinated form of vegetative growth, often with random production of differentiated structures, resulting in loss of normal growth habit.
  2. this mutation appears as randomly occurring growth from multiple growing points and usually results in a misshapen, club-like plant.
  3. abnormally formed.

- **montane**
  occurring in mountains.

- **moor**
  1. an expanse of open, uncultivated and more or less elevated land, often overrun with heath.
  2. (ME. *mor*, fr. OE *mor*; akin MD. *moer*, mire, swamp) chiefly British: an extensive area of open rolling infertile land consisting of sand, rock or peat usually covered with heather, bracken, coarse grass and sphagnum moss, a boggy area of wasteland usually dominated by grasses and sedges growing in a thick layer of peat.

- **morphology**
  1. the science of the form and structure of animals and plants and of their organs.

2. (G. *morphologie*, fr. Gr. *morph* - (fr. *morphe*, form) + G. *-logie*, -logy, more at form) a branch of biology that deals with the form and structure of animals and plants, a study of the forms, relationships, metamorphoses and phylogenetic development of organs apart from their functions.

- **morphometrics**

method to define and describe morphological characters and character states based on quantifiable measurements, morphometrics can also help define form taxa by evaluating ranges of variation within and between groups.

- **morphon**

a group of form-species exhibiting continuous variation of morphological characteristics in a single time sequence ('horizontal' variation.)

- **mosaic**

unusual mottling, veining or streaking, in the leaves of a plant, that is usually very colourful. It is usually caused by a virus.

- **motile**

capable of movement

- **mottled leaf**

a disease caused by a zinc deficiency that reduces the size of leaves and fruits in citrus plants. A disease that causes the leaves of cherry trees to pucker and crinkle.

- **mow**

a pile of hay or sheaves of grain, placed in a barn. To cut down with a sharp tool (i.e. to mow grass with a lawn mower).

- **MS (mass spectrometry)**

a powerful analytical technique used to identify and measure a wide variety of biological and chemical compounds. Applications include molecular weight analysis of proteins and oligosaccharides and the identification and characterization of natural products.

- **mucigel**

1. a slime sheath secreted by roots.
2. a water-soluble gelatinous slime, generally containing modified carbohydrates.
3. a sticky secretion produced on the leaves or other trapping parts of some carnivorous plants to detain (or occasionally lubricate.) prey.

- **mucro**

1. a short sharp point on the tip of a flower or leaf.
2. a short, sharp spur as on a tip of a leaf.

3. (L. *mucro*, sharp point) a stiff or sharp point abruptly terminating an organ. A small awn.

- **mucronate**
  1. ending with a triangular tip or a sharp point.
  2. terminated by a sharp tip.
  3. terminating in a sharp point, more acute than the leaf it projects from.

- **mucronulate**
  having a very small mucro.

- **mulch**
  any substance, usually a layer of organic material, applied to the soil over or around a plant to conserve moisture, protect the roots from frost, reduce the growth of weeds and enrich the soil (i.e. bark chips, rotted manure, garden compost, gravel, black plastic.)

- **mulching paper**
  a biodegradable sheet made of peat that is used to reduce water evaporation and keep weeds down.

- **mule**
  a plant hybrid that is self-sterile and usually cross-sterile due to infertile pollen or undeveloped pistils.

- **mull**
  a humus-rich layer of forested soils consisting of mixed organic and mineral matter.

- **multialxial**
  axis composed of multiple, longitudinal filaments, each derived from one apical cell.

- **multibaculate**
  synonym of pluricolumellate.

- **multicarpellate**
  having a compound pistil or ovary formed by the joining of several carpels.

- **multicellular**
  of an organism composed of many cells.

- **multicellular trichome**
  consists of several cells.

- **multigeniculata**
  with many angled stems.

- **multiplanar tetrad**
  a tetrad in which the individual members are arranged in more than one plane.

- **multiple epidermis**
  an epidermis that is more than one cell layer thick.

- **multiple fruit**
  a fruit such as the pineapple, which is actually the fruits of several flowers that have formed in a single cluster.

- **multiseriate rays**
  rays that are more than one cell wide.

- **multistratose epidermis**
  a multiple epidermis in which all layers are derived from the protoderm.

- **muriate of potash**
  this is a slow-acting fertiliser that adds oxide of potassium to the soil. Same as potassium chloride.

- **muricate**
  1. having a rough surface due to the presence of many tiny sharp points.
  2. (L. *muricatus*, having sharp points) having a rough surface texture owing to small, sharp projections.
  3. roughened by sharp points.

- **murornate sculpture elements**
  elevations of the general surface. Examples, cristae, muri.

- **murus**
  a ridge that is part of the ornamentation and, for example, separates the lumina in a reticulate pollen grain or the striae in striate pollen grain.

- **muscariform**
  brush-shaped, penicillatate.

- **musciform**
  relating to or resembling moss.

- **mutation**
  1. an alteration in genetic material caused by an actual physical change within the genes of a cell. When such changes occur in the germ cells, they are able to be passed on to the offspring as inherited traits, producing such effects as a leaf shape or flower colour different from that of the parent.
  2. a change to the nucleotide sequence in a gene or other DNA sequence. A mutation may be harmful, beneficial or be of no consequence to the plant.

- **muticous**
  not pointed.

- **mycelial**
  pertaining to mycelium.

- **mycelioid**
  resembling mycelium.

- **mycelium**
  the mass of filaments that form the root-like part of fungi.

- **mycology**
  the science of fungi. The study of mushrooms.

- **mycorrhiza**
  1. a symbiotic relationship between certain fungi and the roots of plants, including orchids and many types of tree. Mycorrhiza may be ectotrophic, where the fungal hyphae invades the host roots or endotrophic where the fungal hyphae are entirely within the root cells. The fungal hy-

phae can help the roots to absorb essential nutrients and in some species are essential to the normal growth of the plant.
2. a nutritional relationship between certain fungi and the roots and seedlings of orchids (as well as other plants.)
3. a symbiotic relationship between plant roots and fungi.

- **nannofossil**
fossils of minute planktonic organisms, especially calcareous unicellular algae.

- **nano-**
a prefix for elements smaller than 0.5μm.

- **nanoplankton**
plankton from 2 to 20 microns in diameter that would pass through the mesh of a normal plankton net.

- **napiform**
1. turnip shaped (root).
2. turnip shaped, a depressed sphere.

- **nastic**
of a movement made by a part of a fixed plant in response to a non-directional or diffuse stimulus. The direction of movement is always related to the nature of the plant part and not to the stimulus.

- **natural selection**
the driving force behind evolution, whereby organisms best adapted to their environment and most able to withstand environmental pressures are most likely to survive and reproduce, ensuring propagation and survival of their genes through the gene pool.

- **naturalise**
vt. (Fr. *naturel*, natural) to adapt to an environment not native, of foreign origin, but established and reproducing as though native.

- **naviculate (adj)**
boat-shaped, concave, tapering at each end and keeled.

- **nectar**
1. a sugary secretion produced by flowers and other vegetative parts. Nectar attracts pollinators to flowers.
2. (L. *nectar*, nectar; Gr. *nektar*, the drink of the gods, from base of necros, dead, dead body and tar-, who overcomes, hence, death overcoming. so named because the drink was held to confer immortality) the sweetish liquid in many flowers used by bees for the making of honey.
3. a sweet secretion used to attract insects and sometimes

other animals, often for pollination or consumption.

- **nectary**
  1. a multicellular glandular structure secreting nectar. Found in flowers and on vegetative parts in some species.
  2. a gland that secretes nectar.
  3. (Gr. *nektar*, nectar) a part of a flower that secretes nectar. pl. nectaries.
  4. a specialised gland producing nectar.

- **negative reticulum**
  a general term used to describe patterns of ornamentation in which sexine areas are separated by narrow, reticulately arranged grooves.

- **nemathecium**
  wart-like structure bearing reproductive parts.

- **nematocyst**
  complex ejectile organellee produced in a new dinoflagellate genera.

- **neomorphosis**
  (Gr. *neos*, new; *morphosis*, change) regeneration in cases where the new part is unlike anything in the body.

- **neoteny**
  (Gr. *neos*, young; *teinein*, to extend, stretch) the retention of juvenile characteristics in the adult individual.

- **neotype**
  a specimen selected to serve as the taxonomic species type when the original type material is missing.

- **nepenthes (g)**
  a large genus of over 70 species of usually vine-like carnivorous plant distributed through South East Asia, North Australia and Madagascar, forming urns at the leaf extremities; Asian Pitcher Plant.

- **nephriform**
  kidney-shaped.

- **neri**
  of a species with a large number of a particular differentiated structure from the Greek *nêrithmos* = countless. e.g. countless spines = *nerispina, nerispinus.*

- **neritic**
  living in coastal ocean waters.

- **nervation**
  the pattern of ribs in a leaf.

- **nerve**
  a leaf-rib

- **nerveless**
  lacking nervures.

- **nervure**
  a midrib.

- **net**
  veins uniting to form a network.

- **neuston**
  organisms that live at the ocean-atmosphere interface.

- **neutral**
  1. applied to flowers with neither stamens nor pistils. Applied to soils, liquids etc. which are neither acidic nor basic.
  2. with a pH value of 7, the point at which the soil is neither acid or alkaline.

- **neutral flower**
  said of a sterile flower composed of a perianth without any sexual organs.

- **new world**
  coloquial- the north, central and southern Americas.

- **nexine**
  the inner, non-sculptured part of the exine which lies below the sexine.

- **nexine 1**
  synonym of foot layer, pedium, sole.

- **nexine 2**
  synonym of endexine.

- **niche**
  the position that an organism occupies within its environment.

- **nitrogen fixing**
  of plants, often legumes, which are able to extract and fix nitrogen from the air by means of a symbiotic relationship with bacteria (e.g. *Rhizobium species,*) often in root nodules.

- **nitrogen-fixation**
  incorporation of atmospheric nitrogen as a nitrogen source in the cells of certain bacteria and Cyanobacteria.

- **nitrogenous base**
  see **base**.

- **NMR (nuclear magnetic resonance) spectroscopy**
  a non-invasive analytical technique with an infinitely broad range of applications in plant biotechnology and food processing. It is similar to MRI (Magnetic Resonance Imaging) in that it can be used for medical applications. Non-destructive analysis with this tool can be applied to a wide range of research challenges, including the determination of chemical structure, microimaging and plant metabolism studies.

- **nocturnal**
  belonging to the night, as of flowers that open at night.

• **noddling**
describing flowers which hang from a stalk, facing downwards - eg. Sarracenia

• **node**
1. the place on a stem where leaves or branches originate.
2. region of a stem where one or more leaves are attached. compare to internode.
3. the branching points on a cladogram, which are supported by synapomorphies.
4. points on a stem from which leaves, shoots or flowers grow. The word 'joint' means the same thing.

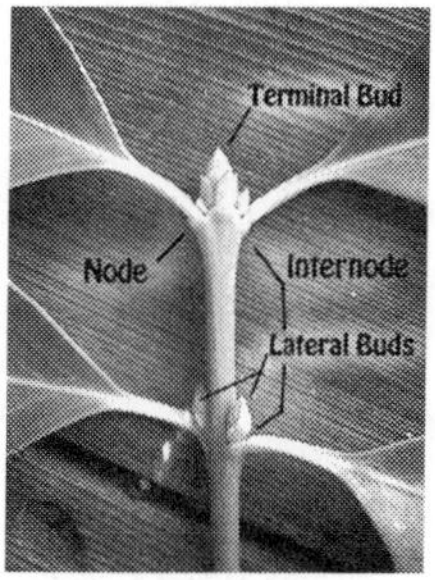

5. point on the stem where leaves are attached or the point of branching of the stem.
6. site on an axis from which leaves or branches arise.
7. (L. *nodus*, knob) a knob or joint of a stem from which leaves, roots, shoots or flowers may arise. A node will contain one or more buds.

• **nodose**
nodular, knotty.

• **nomen conservandum**
conserved name- the name of a taxon, usually one in which the name is in common usage, which is internationally agreed to be exempt from the usual rules of priority of publication.

• **nomen nudum**
naked name- a name published without an accompanying description.

• **nomenclature**
the study of the systematic naming of taxa, using Latin names so that they will be intelligible and uniform in all countries.

• **nomenclature**
(L. *nomen*, name; *calare*, to call) the making and giving distinguishing names to all groups of plants.

• **non-aperturate**
synonym of inaperturate.

• **non-resupinate**
referring to Orchid flowers which don't twist 180 degrees during development or which twist a full 360 degrees, with the

result that the *labellum* is uppermost.

- **non-storied cambium**
  fusiform initials not aligned with one another.

- **normal**
  typical of its taxon, of features held in common between specimens in the taxon.

- **normapolles**
  a group of Cretaceous and Lower Palaeogene pollen, usually triporate, with a complex pore apparatus.

- **novel foods**
  products that have never been used as a food; foods which result from a process that has not previously been used for food or foods that have been modified by genetic manipulation (genetically modified foods.)

- **npc-classification**
  a morphological system for classifying pollen grains and spores that was based on the number, position and character of their apertures.

- **nucellus**
  1. the central part of the ovule enclosing the female gametophyte.
  2. ovule tissue within which an embryo develops (embryo sac); homologous with the megasporangium of a seed plant.
  3. the inner nutritive layer of the ovule.

- **nucleolus**
  a small round densely-staining region within the cell nucleus which contains ribosomal rRNA subunits and nuclear organisers (multiple copies of genes encoding rRNA).

- **nucleotide**
  the building blocks of DNA which consist of a base, sugar and phosphate group.

- **nucleus**
  a large organelle, usually located near the center of the cell and containing the DNA.

- **nuda**
  naked, without spines, wool etc.

- **nudate**
  synonym of psilate.

- **nudiflorus**
  naked flowers without scales, spines or wool on the ovary and flower tube.

- **numerical taxonomy**
  same as phenetics a method of generating phylogenies that is based on large numbers of quantifiable (measureable) characters which groups organisms with respect to overall similarity.

- **nurse plant**
  a plant, often a shrub, which provides shelter and shade to developing seedlings of other plants.

- **nut**
  1. a hard-walled fruit containing a single seed that does not split spontaneously.
  2. (ME. *nute, note*, fr. OE *hnute*; akin to OHG *nuz, hnuz*, nut) a hard-shelled dry fruit or seed having a more or less distinct separatable rind or shell and interior kernel or meat; a dry indehiscent one-seeded fruit with a woody pericarp developing from an inferior syncarpous ovary.

- **nutlet**
  a small nut.

- **nutraceutical**
  a product isolated or purified from food that is generally sold in medicinal forms not usually associated with food. A nutra-ceutical has been demonstrated to have a physiological benefit or provide protection against chronic disease.

- **nutrient**
  a chemical substance that an organism ingests to enable it to survive e.g. essential minerals.

- **nutritive cell**
  specific cell of carpogonial branch with which carpogonium fuses after fertilisation in some red algae of the order Cryptonemiales.

- **nyctinasty**
  of a movement in response to changing day and night conditions.

- **ob**
  a prefix to another structural descriptor of a part, indicating attachment in the opposite or inverse way to that expected from the rest of the descriptor. Hence *obovate* / ovate but attached at the narrow end.

- **obconate**
  conical with the attachment at the apex of the cone.

- **obcordate**
  inversely heart-shaped. The wider end and its deep notch at the top instead of the bottom.

- **obdeltoid**
  inversely deltoid. Triangular, with the apex downward.

- **obdiplostemonous**
  having the stamens in two whorls, those of the outer whorl being opposite the petals.

- **obdiplostemony**
  the condition of being obdiplopstemonous.

- **oblanceolate**
  1. lanceolate but attached at the narrow end.
  2. inversely lanceolate, shaped like a lance-point reversed, with the tapering point attached to the leafstalk.
  3. (L. *ob*, reversely; *lancea*, spear) shaped like a lance point reversed, that is, having the tapering point next to the leafstalk.

- **oblate**
  describing the shape of a pollen grain or spore in which the polar axis is shorter than the equatorial diameter. See **oblate spheroidal, P/E ratio, peroblate, perprolate, prolate, prolate spheroidal, shape classes, spheroidal, suboblate, subprolate, subspheroidal.**

- **oblate spheroidal**
  describing the shape of a pollen grain or spore in which the ratio between the polar axis and the equatorial diameter is 0.88-1.00. See **oblate, P/E ratio, peroblate, perprolate, prolate, prolate spheroidal, shape classes, spheroidal, suboblate, subprolate, subspheroidal.**

- **obligate**
  1. compulsory, essential, unable to exist under alternative conditions.
  2. restricted to a particular way of life.

- **oblique**
  1. in a diagonal position.
  2. slanting with unequal sides.
  3. (L. *obliquus*, slanting) slanting; unequal-sided.
  4. (usually of a leaf) having unequal sides; almost (but not quite,) symmetrical.

- **oblong**
  1. longer than broad and mostly parallel-sided, such as an oblong leaf.
  2. (L. *oblongus*, rather long) elliptical and from two to four times as long as broad.
  3. broadly elliptical with blunt ends.

- **obovate**
  1. ovate but attached at the narrow end.
  2. ovoid but attached at the narrow end.
  3. inversely ovate; having the broad end upward or toward the apex, as in many leaves.
  4. (L. *ob*, against; *ovum*, egg) inversely ovate, having the shape of the longitudinal section of an egg, with the broad end at the top, as some leaves.
  5. ovate with the narrow end at the base.

- **obovoid**

(L. *ob*, against; *ovum*, egg; Gr. *eidos*, shape) inversely ovoid; roughly egg-shaped, with narrow end downwards, said of some fruits.

- **obrotund**

approaching a round form.

- **obsolete**

(L. *obsolescere*, to go out of use) not developed, rudimentary or not evident. Applied to a structure that is almost suppressed.

- **obtuse**

1. with an angle greater than 90 degrees and less than 180 degrees, blunt.
2. blunt or rounded at the extremity, as an obtuse leaf, sepal or petal.
3. blunt or rounded at the end.
4. (L. *obtusus*, blunt) with blunt or rounded end.
5. terminating in a blunt, unpointed end, forming an angle of more than 90 degrees.

- **ocean**

(ME. *ocean*; L. *oceanus*, fr. Gr. *okeanos*, the ocean) the great body of salt water that covers mores than two thirds of the surface of the earth; any of its five principal geographical divisions, the Atlantic, Pacific, Indian, Arctic and Antartic.

- **ocrea**

1. a sheathing stipule or a pair of stipules united into a tubular sheath around the stem.
2. (L. *acrea*, greave or legging) a tubelike covering around some stems, especially of plants of the Polygonaceae.

- **oculus / oculi / oculate**

the much enlarged part of the pore structure in pollen of the fossil Normapolles group. Example, Oculopollis.

- **odd-pinnate**

1. pinnate leaves (leaves w/ opposite pairs) with a terminal odd leaflet, as in the rose.
2. said of compound leaves having an odd number of leaflets, this is usually easily determined because there is a single terminal leaflet.

- **odoriferous**
giving odor or scent, usually sweet; diffusing fragrance, fragrant, as odoriferous flowers.

- **offset**
1. a plant or plants growing from a side-shoot, rhizome or tuber that remains connected to the main plant, often easily removed for propagation.
2. a young plant produced by the parent, usually as its base also called offshoot. A small bulb at the base of a mother bulb.
3. to produce offsets.

- **oil cavities**
a cavity lined with cells that secrete oils.

- **oilseeds**
seeds that are used in the production of cooking oils and other food products, such as margarine. Oilseeds include canola, soybean, flaxseed, safflower and sunflower seeds.

- **oligotrophic**
nutrient-deficient.

- **olliform**
in the shape of a flower-pot. A truncated cone.

- **ol-pattern**
a pattern of ornamentation that appears to show 'dark islands' at high focus (H) and that become brighı at low focus (L).

- **omniaperturate**
describing a pollen grain in which the exine is very thin or absent and the intine is thick, so that no specific apertural region can be distinguished and thus the whole surface can be considered apertural in nature.

- **omnivore**
an animal that eats both plants and animals.

- **oncus**
a lens-shaped structure that is not resistant to acetolysis and occurs beneath the apertures of many kinds of pollen grains. Example, Corylus (Betulaceae)

- **ontogeny**
1. sequential development of an individual through all its stages.
2. the course of development of an individual organism.

- **oogamous**
being reproduced by the fusion of a small, mobile male gamete with a large, immobile female gamete.

- **oogamy**
1. reproduction in which the female gamete is large and non-motile and the male gamete small and motile.
2. union of gametes in which a sperm fertilises a relatively large, non-motile egg.

- **oogonium**

female reproductive cell containing one or more female gametes or eggs.

- **open**

no distinct boundary separates the root cap from the root proper.

- **open-pollinated**

pollinated by wind or insects, not by human intervention.

- **operculum**

1. the lid of a pitcher or similar.
2. a distinctly delimited sexinee-ctexine structure which covers part of an ectoaperture and which is completely isolated from the rest of the sexine.

- **opposite**

1. parts of plant, such as leaves, arranged in pairs on opposite sides of the stem.
2. leaves, which are arranged in pairs. Two leaves grow from one *node* directly across from each other on each side of a stem.
3. (L. *opponere*, to oppose) said of leaves or bracts occurring two at a node on opposite sides of the stem. Said of flower parts when one part occurs in front of another.

- **optical (cross-) section**

the image seen in optical microscopy when the plane of focus is half way through a palynomorph.

- **orbicular**

1. round and flat or spheroidal.
2. nearly or completely circular, as an orbicular leaf. Also orbiculate.
3. (L. *orbis*, circle) round or shield-shaped with petiole attached to centre.
4. disc-shaped; perfectly circular.
5. (usually of a leaf) almost circular.
6. a general term, applied in palynology for an orbicular granule of sporopollenin -sporopollenin.

- **orchard**

a garden. A piece of ground, usually enclosed, devoted to the cultivation of fruit trees, especially apple, pear, peach, plum and cherry trees. A collection of cultivated fruit trees.

- **order**

or evolutionary order; which character state that must follow another in a character transformation series - without any implication as to what characters are primitive versus derived. For example, the order of the character states 'grey', 'white' and 'black' could be hypothesised to be white-grey-black (or black-grey-white.)

- **order**

  1. the taxonomic grouping between class and family.
  2. a group of related families taxonomically classified within a class.

- **ordinary epidermal cells**

  the least specialised cells of the epidermis (i.e. cells that are NOT specialised as guard cells, root hairs, trichomes etc.). May secrete a cuticle.

- **ordinate**

  describing a pollen grain or spore with an arrangement of elements regularly distributed.

- **organ**

  1. a distinct and visibly differentiated part of a plant such as a leaf.
  2. a specialised plant structure performing a particular function.

- **organelle**

  1. a distinct sub-cellular structure, usually with a specific function and compartmentalised within its own membrane.
  2. (literally 'little organ') a region within a cell where specialised metabolic tasks occur, typically surrounded by a membrane.
  3. specialised structure in a cell, like a plastid or nucleus.

- **organic**

  1. of a mainly carbon-based molecule or chemical or chemistry.
  2. a style of agriculture where synthetic pesticides and fertilisers are not used.
  3. of food produced by organic farming methods.
  4. compounds containing carbon derived from decomposed animal or plant organisms. Is also used loosely of mulches, soil mix etc. taken from plant materials.

- **organic gardening**

  a method of gardening using only organic materials, such as manure, peat moss and compost and rejecting inorganic material, such as rock phosphate, ammonium, insecticides and pesticides.

- **organism**

  a complete, living individual member of a species. Any individual living creature, either multicellular or unicellular.

- **ornamentation**
  a general term that is useful for describing the organisation of features.

- **ornate**
  describing a reticulate ornamentation consisting of broad, curved muri and lumina that are often anastomosing. Example, Ceiba aesculifolia (Bombacaceae).

- **orthocolpate**
  describing a pollen grain with the colpi in the most common position, perpendicular to the equator.

- **orthostichous**
  exhibiting orthostichy, arranged in vertical ranks.

- **orthostichy**
  a vertical rank; an arrangement of leaves or scales at different heights on an axis so that their median planes coincide, as the vertical ranks of leaves on a stem.

- **orthotropic**
  of or pertaining to or exhibiting orthotropism; growing more or less vertically.

- **orthotropism**
  vertical growth, the tendency of plants to grow more or less vertically.

- **orthotropous**
  a term describing a condition in which all parts of a seed grow in a straight line.

- **osmophore**
  a gland, usually floral, which produces fragrances that attract pollinators.

- **osmosis**
  1. the movement of water through a selectively permeable membrane from a place of low solute concentration to a place of higher solute concentration.

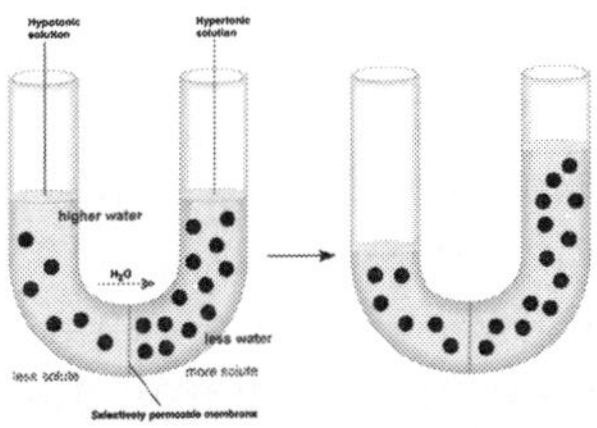

  2. the diffusion of fluids through a semipermeable membrane, as in a living cell.

- **osteosclereids**
  bone shaped, elongated with swollen ends.

- **ostiole**
  opening in conceptacle or cystocarp through which spores or gametes can escape.

- **OTU**
  Operational Taxonomic Unit. definitions or names of the taxa

included in a phylogenetic analysis.

- **outcrossing**

  the transfer of genetic information among plants, generally through pollen. Outcrossing occurs for all plants whether genetically modified or not.

- **outline**

  a general descriptive word. Applied in descriptive terms like equatorial outline and outline in polar view.

- **oval**

  elliptical

- **ovary**

  1. the female part of the plant, below the style, containing the egg cells (ovules) that after fertilisation enlarge and become seeds.
  2. the enlarged basal portion of a carpel, where the ovules are borne. The ovary differentiates into the fruit.
  3. the enlarged, rounded, *ovule*-producing base of the female portion (the *pistil*) of a flower that eventually develops into a fruit after being fertilised.
  4. (L. *ovum*, an egg) the enlarged hollow part of a pistil in angiosperms in which ovules are formed.
  5. a hollow organ at the base of the carpel of a flower in which the female plant's gametes are produced.

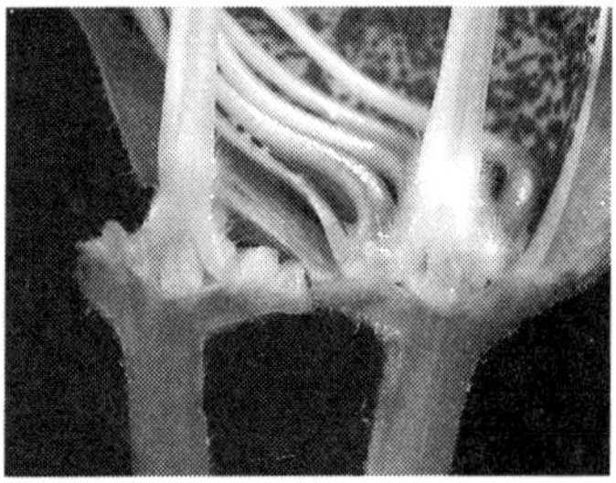

- **ovate**

  1. a flattened structure (e.g. leaf) having an outline in the shape of an egg, with the stalk or point of attachment at the large end.
  2. (L. *ovum*, an egg) having the shape of a longitudinal section of an egg, egg-shaped and attached by the broader end.
  3. egg-shaped, oval, with one end more pointed than the other.
  4. the leaf is egg-shaped, the broadest part is below the middle.

- **overall similarity**

  a method by which organisms that share the most similarities are grouped together; characters are not distinguished as to whether they are primitive or derived or whether they are evolutionary meaningful.

- **overarched**
  forming an arch over.
- **oviparous**
  (L. *ovum*, an egg; *parere*, to produce) egg-laying; producing eggs which hatch after leaving the body of the female. Germinating while still attached to the parent plant, for example, mangrove.
- **ovoid**
  1. a structure with a circular cross section (i.e. not flattened as in ovate) having an outline in the shape of an egg, with the stalk or point of attachment at the large end.
  2. (L. *ovum*, an egg) egg-shaped.
- **ovoidal**
  egg-shaped, ovate.
- **ovule**
  1. the outgrowth in a seed-plant's ovary that develops into a seed following fertilisation of the egg cell within it. The fertilised egg cell forms a group of cells that becomes an embryo consisting of an axis, (root and stem or hypocotyl and epicotyl) and in *Cactaceae* two cotyledons, with its associated protective and nutritive structures capable of developing into a plant.
  2. unfertilised seed, the ovule contains the megasporangium with the megagametophyte, surrounded by one or two integuments integument.
  3. an outgrowth of the *ovary*, which upon fertilisation, becomes the seed. An embryo sac with egg surrounded by nucellus and two integuments.
  4. (L. *ovum*, an egg) a structure in seed plants which contains the megasporangium (nucellus), megaspore (embryo sac), a food store and a coat and develops into a seed after fertilisation.
  5. the part of the ovary of seed plants that contains the gamete. an unfertilised seed.
- **oxidation**
  a chemical change involving the addition of oxygen.
- **oxide**
  a compound of any element mixed with oxygen alone.
- **oxygen**
  an element found in the soil needed by plants for growth.
- **P/E ratio**
  the ratio of the length of the polar axis (P) to the equatorial diameter (E).
- **palaeopalynology**
  the study of fossil palynomorphs.
- **pale**
  1. the palea of a grass.

2. a membrane dividing distinct parts of a plant structure

- **palea**

1. one of the chaff-like bracts or scales under the individual flowers in the heads of many Compositae (Daisy family), as in the sunflower.
2. the upper bract that encloses, with the lemma, the flower in grasses. Also palet.
3. the scales on the stems of certain ferns, ramentum.
4. (L. *palea*, chaff) the upper and usually shorter and thinner, of two membranous bracts enclosing the flower in grasses.
5. a chaff-like bract usually found in grasses.

- **paleaceous**

chaffy or chaff-like, covered with chaffy scales.

- **paleate**

clothed with scales.

- **paleobotany**

the study of plants that existed in former geological periods, mainly known from their fossils or pollen.

- **paleopalynology**

the study of fossil palynomorphs, where 'fossil' is defined as 'any indication of past life' including the term 'sub-fossil'. Fossil palynomorphs are comprised of sporopollenin, chitin or pseudochitin little modified from their original composition. Specific applications include stratigraphic palynology, archaeological palynology and environmental palynology.

- **palisade mesophyll**

the region of ground tissue in a leaf where the chlorenchyma cells are elongated and arranged perpendicular to the epidermis, usually in the upper half of the leaf.

- **palm**

a monocotyledon of the family Palmae with an unbranched trunk or stem crowned with large pinnate (fan-shaped) leaves. Most species are tropical or sub-tropical and some are economically important.

- **palmate**

1. divided with radial lobes like the fingers of a hand.
2. having four or more lobes, leaflets or nerves radiating from a single point, as in the leaves of the horse chestnut (Aesculus hippocastanum). The leaves are

arranged as a hand with outstretched fingers.

3. radiately lobed or divided.
4. (L. *palma*, palm) leaves divided into lobes arising from a common centre.
5. 5-lobed, with the midribs joined at a common point.

- **palmatifid**
divided in a palmate manner about halfway to the base.

- **palmatisect**
cleft in a palmate manner, usually no more than halfway to the base.

- **palynodebris**
all palynomorph-sized particles in a sediment excluding those that actually are palynomorphs but including, for example, wood fragments, cuticles and some animal remains.

- **palynodeme**
a group of palynomorph species (form-species) that intergrade and represent the reflection of a known or hypothetical plant species.

- **palynofacies**
the assemblage of phytoclasts found in a particular sediment, such as palynomorphs, wood fragments, cuticles etc.

- **palynogram**
a diagram summarising the main morphological features of a palynomorph.

- **palynology**
the study of pollen grains and spores and of other biological materials that can be studied by means of palynological techniques.

- **palynomorph**
fossil spores and pollen; also dinoflagellates.

- **palynomorph**
a general term for all entities found in palynological preparations.

- **pandurate**
in the shape of a violin, an oval figure with sides pinched in to produce 2 lobes, one larger than the other.

- **panduriform**
in the shape of a violin, an oval figure with sides pinched in to produce 2 lobes, one larger than the other.

- **panicle**
1. a branched flower cluster.

2. a loosely branched, pyramidal flower cluster.
3. (L. *panicula*, a tuft of plants) a branched racemose inflorescence often applied more widely to any branched inflorescence.
4. a branching cluster of flowers esp. where formed by a series of racemes branching from a main axis.

- **panicled**
furnished with or arranged in branched clusters.

- **paniculate**
arranged in branched clusters. Also panicled.

- **paniculate**
panicled, arranged or growing in panicles.

- **pannose**
looking or feeling like felt or wool cloth.

- **panporate**
synonym of pantoporate.

- **panto-, pan-**
a prefix for global distribution.

- **pantoaperturate**
describing a pollen grain with apertures spread over the surface sometimes forming a regular pattern.

- **papilla**
1. small, rounded protruberance from any plant surface. Hence Adj. *Papillose* - of or relating to or resembling papilla.
2. (L. *papilla*, nipple) a glandular hair with one secreting cell above the epidermis level.
3. a general term, applied in palynology to parallel sided exinous elements with rounded apices, less than 1μm in length.
4. a small protuberance.

- **papillae**
a small, nipple-like protuberance. A rounded projection from a portion of an epidermal cell.

- **papillary**
pertaining to or resembling a papilla. Also papillar; papillose, papilate.

- **papilliferous**
bearing nipple-shaped protuberances.

- **papilliform**
resembling a nipple-shaped protuberance.

- **papillose**
(L. *papilla*, nipple) descriptive of a surface beset with short, blunt, rounded or cylindric projections.

- **pappose**
describes a plant having or resembling the type of bristly or feathery perianth of composite plants such as thistles and dan-

delions. Also pappous.

- **pappus**
a bristly, feathery or fluffy perianth whorl crowning the ovary or fruit in the Compositae, adapted for dispersal of the fruit by the wind or other means. A good example of a plant having a pappus is the Dandelion.

- **papyracantha**
with papery spines.

- **papyraceous**
made of or resembling papyrus or paper.

- **parabiosis**
the natural or artificial joining or grafting of two organisms.

- **parabolical**
a more rounded form of rhomboid .

- **paracavate**
an exine in which the intexine is clearly defined but in which its degree of separation from the exoexine is uncertain or indeterminate. Example, Ancyrospora langii.

- **paraisopolar**
describing a pollen grain whose polar faces differ only in the attachment of viscin threads to the proximal pole.

- **parallel-veined**
having veins quite or nearly parallel, as the leaves of Tibouchina urvilleana.

- **paramylon**
a glucose polymer which serves a food reserve in Euglenophyta.

- **paraphyletic**
of a group of organisms sharing and including a common ancestor, but not including all the descendents of that common ancestor.

- **paraphyly, paraphyletic group**
terms applied to a group of organisms that include an ancestor and some, but not all of its descendants, compare to monophyly, monophyletic groups and polyphyly and polyphyletic groups.

- **paraphysate**
having or marked b paraphyses.

- **paraphyses**
hairs or hair-like structures in the sorus.

- **paraphysis**
1. one of the erect, slender, sterile filaments accompanying the spore-bearing or sexual organs of spore-bearing plants such as ferns.
2. a sterile hairlike structure among sporangia or gametan-

gia, as those in the brown algal order Fucales.

- **paraporal lacuna**
  a lacuna of a lophate pollen grain lying in the mesocolpial region adjacent to one side of an equatorial ridge. Example, Taraxacum officinale (Compositae).

- **paraporal ridge**
  a ridge bounding a pore, extending in a meridional direction. Applied to a lophate pollen grain. Examples, Tragopogon pratensis, Taraxacum officinale (Composiatae).

- **parasite**
  1. an organism that grows on or within another, drawing its nutrients from it e.g. mistletoe.
  2. an organism that grows in or on another plant or animal, getting all or part of its nutrients from it and usually showing a degree of adaptive structural modification.

- **parastichy**
  reference to the virtual spirals in plants, joining elements (tubercle, leaf, scale, floret) to their nearest neighbours. Parastichies usually come in two families winding in opposite directions. Parastichy numbers denoted by a pair (n,m) classify spiral and whorl phyllotaxes.

- **parasyncolpate and parasyncolporate**
  describing syncolp(or)ate pollen grains in which the apices of the colpi divide into two branches and anastomose towards the poles, delimiting an isolated area known as the apocolpial field. Examples, Nymphoides peltata (Menyanthaceae), Eugenia (Myrtaceae).

- **paratype**
  additional specimens, other the holotype, cited at the same time as the original description.

- **parenchyma**
  living undifferentiated cellular tissue with thin cell walls, which may differentiate to form other types of cell. Parenchyma cells are the most common plant cells, make and store food substances and form the pith at the centre of stems of dicotyledons.

- **parenchymal**
  pertaining to or of the nature of parenchyma.

- **parichnos**
  an interconnected system of parenchymatous strands with many air spaces that extend throughout the vegetative organs of some arborescent lycopsids.

• **parichnos scars**
small scars that marks the position of the parichnos strands on stem fossils of the some arborescent lycopsids.

• **parietal**
1. borne on or pertaining to the wall or inner surface of a capsule or other hollow structure. Of ovules arising from the outer wall of the ovary and pointing inwards.
2. (L. *paries*, wall) when the placenta is attached to the wall of the ovary.
3. of the wall of a hollow stem or other structure borne on the inner surface of a capsule.

• **parted**
1. divided nearly to the base, as in leaves. Also partite.
2. lobed, but only joined by the base.

• **parthenogenesis**
the reproduction of offspring from an egg cell without fertilisation.

• **parthenogenetic**
production of new individual from single, unfertilised gamete, usually the egg.

• **partial dimorphism**
leaves with only a part modified as fertile sporangia.

• **parviflorus**
with small flowers.

• **passive trap**
a trap which is incapable of movement to catch or detain its prey

• **patelliform**
kneecap shaped, broadly round, convex on one side and concave on the other.

• **pathogen**
1. any micro-organism which causes disease, especially bacteria, fungi and viruses.
2. an organism, usually a microbe, that causes adverse effects in its host that may cause disease.

• **patina / patinate**
a thickening of the exine of spores that extends over the entire surface of one hemisphere. Example, Cingulatisporites, Patellasporites.

• **pattern**
a general word, applied in palynology either to surface features or infratectal elements, such as columellae.

• **paucispinus**
with few spines.

• **PCR (polymerase chain reaction)**
a method of increasing the quantity of a specific DNA sequence.

- **peat**

(ME. *pete*, fr. ML. *peta*, perh. of Celt. Origin, akin to W. *peth*, thing - more at piece) a piece of turf cut for use as a fuel, a mass of partially carbonised plant tissue formed by partial decomposition in water of various plants and esp. of mosses of the genus Sphagnum, widely found in many parts of the world, varying in consistency from a turf to a slime used as a fertiliser, as stable litter, as a fuel and for making charcoal.

- **pectin**

1. complex carbohydrates derived from poly-galacturonic acid, present in inter cellular substances (middle lamella and cell walls. Pectins may be divided into the three) general types- protopectin, pectin and pectic acid.
2. the hot-water-soluble portion of the cell wall matrix.

- **pectinate**

1. comb-like- often used to refer to species in which the spines (or other structures) are spread out like the teeth of a comb.
2. resembling a comb, divided nearly to the base with narrow, close segments, as the leaves of some ferns.
3. pinnatifid with closely set segments, comb-like.
4. (L. *pecten*, comb) comb-like.

- **pedate**

1. palmate, with the two lateral lobed divided into smaller segments, such as the leaf of Helleborus foetidus.
2. palmately cut or divided with the lower pair basiscopically exaggerated.
3. palmate, but with the first and fifth lobes tridentate, to give a 7-fid leaf overall.

- **pedicel**

1. the short stalk connecting a flower to the main stem of a plant or branches of its inflorescence.
2. the stalk of a flower. Also pedicle.
3. (L. *pedicellus*, foot) the stalk of a flower in an inflorescence.
4. any small, stalk-like structure.

- **pedicellate**

having or attached by a pedicel or pedicels.

- **pedicelliform**

having the form of a pedicel.

- **peduncle**

1. a botanical term used to describe the stalk of a single flower, as in the Tulip or of a cluster of flowers, as in the Geranium.

2. (LL. *pedunculus*, small foot) the stalk of a flower borne singly or the stalk of an inflorescence.
3. the stalk of a flower, inflorescence or fruit.

- **pelagic**
living in the open ocean, offshore.

- **peltate**
(Gr. *pelta*, target) shield-shaped, leaves that are shaped like a shield and attached to the stem at the center or by some point distinctly within the margin and having the petiole inserted into the undersurface of the lamina not far from the centre.

- **peltate (of leaves)**
a leaf in which the leafstalk is connected to the centre of the leaf.

- **pendent**
1. hanging down from a support.
2. hanging or supported from above, drooping. Also pendulous.

- **pendulous**
drooping or hanging down.

- **penicillatate**
brush-shaped.

- **penicillate**
like a tuft of hairs.Having or forming small tufts.

- **peniculllate**
(L. *penicillus*, a pencil or small brush) having the form of a pencil.

- **pennate**
tapering.

- **pentadelphous**
grouped together in five sets, as the stamens of the linden, which are in bunches of five united by their filaments.

- **pentagonous**
with five angles.

- **pentamerous**
consisting of five sets of parts, as in flowers such as many of the rose family, which have calyx lobes, petals, stamens and other flower parts in five's or multiples of five.

- **pentandrous**
pertaining to a flower having five stamens.

- **pentastichous**
describing an arrangement of leaves upon the stem in five vertical rows or ranks, as in the apple tree.

- **peony form**
describes a deep, rounded flower consisting of a convex mass of petals, petaloids and sometimes stamens.

- **percurrent**
running through the entire length from base to tip, as the midrib of a leaf.

- **perennate**
to live perennially from one year to another.

- **perennation**
1. the tendency of a plant to live for more than two growing seasons.
2. (L. *perennis*, perennial) survival of a plant for a number of years. To live over from season to season.

- **perennial**
1. growing for more than two years. *Cactaceae* are generally regarded as perrenials.
2. living for at least three seasons, commonly used of *herbaceous* plants and *woody* perennials, i.e. trees and shrubs.
3. (L. *perennis*, through, *annus*, a year) a plant that grows for 3 or more years and usually flowers each year.

- **perfect**
1. of an organism capable of sexual reproduction. Of a flower with both male and female flower parts.
2. flowers having both *stamens* and *pistils*, also refers to a whole plant whose flowers have both male and female parts.

- **perfect flower**
a flower with both essential and accessory organs.

- **perfoliate**
1. having a stem that seems to pass through the blade, said of a leaf, such as those of many honeysuckles.
2. (L. *per*, through; *folium*, a leaf) said of opposite or whorled leaves or bracts that are united into a collarlike structure around the stem that bears them.

- **perforate**
a general adjective indicating the presence of holes, applied in palynology to holes less than 1μm in diameter and generally situated in the tectum.

- **perforation plate**
the end wall of a vessel element where the secondary cell wall was not deposited and the primary cell wall has been digested.

- **peri-**
synonym of panto-.

- **perianth**
1. the combined structure comprising the calyx (sepals) and corolla (petals). In *Cactaceae*, the sepals and petals form an intergrading series known variously as perianth segments, tepals, sepaloids and petaloids.

1. the term for that part of a flower which surrounds the *pistil* and *stamens*; it generally consists of *sepals* and petals.
2. a collective term for the sepals and petals of the flower, especially when these are indistinguishable.
3. (Gr. *peri*, around, *anthos*, flower) the outer whorl of floral leaves of a flower, when not clearly divided into calyx and corolla, collectively, the calyx and corolla or either one if one is absent.

- **perianth segment (tepal)**
  the leaf like and sterile parts of a flower, especially used if these parts are not distinguishable into calyx and corolla.

- **pericarp**
  1. the wall of a fruit that is fleshy, as in berries or hard, as in nuts.
  2. sterile jacket of cells surrounding the carposporophyte in the Florideophycidae.
  3. (Gr. *peri*, around *karpos*, fruit) the fruit wall which has developed from the ovary wall sometimes used for any fruit covering.
  4. the part of a fruit formed from the wall of a ripened ovary.

- **pericarpial**
  of or relating to the wall of a fruit.

- **pericarpoidal**
  resembling a pericarp.

- **periclinal**
  parallel to the surface.

- **pericycle**
  in roots, the layer of cells between vascular tissue and endodermis that gives rise to lateral roots and vascular cambium.

- **periderm**
  1. a tissue primarily consisting of cork cells,outer bark.
  2. the cork cambium and the tissues in produces, outer bark.

- **perignous**
  surrounding the pistil of a flower. Perianth.

- **perigynium**

1. the sac-like bract that surrounds the pistil in the sedges of the group Carex.
2. (Gr. *peri*, around *gyne*, female) fruit investing utricle of the sedges, Carex.

- **perigynous**

1. sepals, petals and stamens attached to a hypanthium.
2. (Gr. *peri*, around *gyne*, female) growing in a ring around the pistil, as the stamens; having stamens etc. growing in this way, said of a flower.

3. (of stamens) surrounding the pistil.

- **perigyny**

the state or condition of being perigynous.

- **perine**

a sporoderm layer that is not always acetolysis resistant and is situated around the exine of many spores. Example Pteridium (Hypolepidaceae.)

- **period**

jurassic - ca. 213 - 144 million years ago; Cretaceous - ca. 144 - 65 million years ago.

- **perisperm**

1. the nutritional tissue contained in a seed, especially that portion which is formed outside of the embryo sac from the nucellus.
2. the nutritive portion of the endosperm of some seeds, that lies outside the embryo-sac.

- **perispore or perine**

an outer covering of some fern spores, with different configuration than that of the exospore.

- **peristome**

a fringe of small teeth around the mouth of many nepenthes.

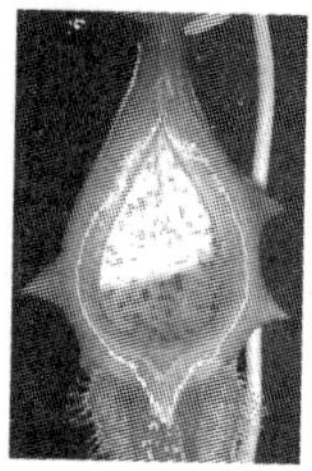

- **peritreme**

describing a pollen grain with equatorial apertures situated around an outline that is circular in polar view.

- **permineralisation**

a preservation type in which mineral matter has infilled intercellular and intracellular spaces, but has not replaced the cell walls compare to petrifaction and silification.

- **peroblate**

describing the shape of a pollen grain or spore in which the ratio between the polar axis and the equatorial diameter is less than 0.50.

- **perprolate**

describing the shape of a pollen grain or spore in which the ratio between the polar axis and the equatorial diameter is more than 2.

- **per-reticulate**

structural elements fused distally forming an open reticulum.

- **persistant**

1. remaining attached instead of falling away.
2. leaves or flower petals that remain attached to the plant instead of dropping off.
3. (L. *persistere*, to persist) remaining attached after the normal function has been completed.
4. (of leaves etc,) remaining on the plant when dead instead of falling off.

- **personate**

of a monopetalous corolla where the lip of which is unequally divided, appearing in the form of a gaping animal.

- **pest**

any very destructive or mischievous animal or insect in a garden.

- **pest netting**

a lightweight net draped over fruit trees, berries or row crops to protect them from rabbits, deer and birds.

- **pest-free plants**

plants resistant to insects.

- **pesticide**

1. a chemical or other substance used to kill pests. Depending on the target species pesticides may be referred to as insecticides (insects), herbicides (plants), fungicides (fungi) and nematicides (nematodes). Pesticides may act on contact or require ingestion.
2. any substance used to kill, repel, prevent or alleviate a pest problem. They include fungicides, insecticides, herbicides and bactericides.

- **petal**

1. one of the segments of the inner perianth segments or corolla. In the *Cactaceae* these are actually petaloid parts of the perianth and not true petals. Petals are often brightly coloured and interior to the sepals.
2. one of the whorls of a flower, petals may be brightly coloured.

3. one portion of the often bright and coloured part of the *corolla* one of the floral leaves which combine to form the perianth of a flower. The whorl of flower parts just inside the sepals, usually colourful. The median petal of Orchids, the *labellum* is often differentiated in form and/or function.
4. interior to sepals, usually conspicuous to attract pollinators.
5. (Gr. *petalon*, leaf) any of the component parts or leaves, of a corolla the unit of structure of the corolla.
6. the floral modified leaves of a flower that comprise the corolla.

- **petaloid**

1. a term meaning petal-like.
2. (Gr. *petalon*, leaf *eidos*, form) like a petal.
3. resembling a petal in colour or texture.
4. (L. *petiolus*, small foot) growing on or provided with, a petiole.
5. relating to or in the form of a petiole.

- **petiole**

1. the leaf stalk.
2. (L. *petiolus*, small foot) the slender stalk or stem of a leaf, also called a leaf stalk.
3. the usually slender stalk of a leaf.

- **petiolule**

the stalk of a leaflet of a compound leaf.

- **petrifaction**

a type of preservation in which mineral matter infills intercellular spaces and replaces the cell walls, compare to permineralisation and silification.

- **pgprs (plant growth promoting rhizobacteria)**

a class of bacteria that enhance canola plant growth by preventing deleterious effects of other microorganisms that are harmful to plants.

- **pH**

a numerical designation of the acidity and alkalinity in soil and other biological systems. A pH of 7.0 indicates neutrality; above 7.0 indicates alkalinity and lower indicates acidity.

- **phaeacantha**

with dusky spines.

- **phagotrophy**

endocytosis or engulfing particles of food as a mode of nutrition.

- **phanerogam**

1. a seed plant or flowering plant, such as a pine tree, poppy

or bamboo, as opposed to a cryptogam or spore-bearing plant, such as a fern or mushroom.

2. the seed plant - gymnosperms or angiosperm.The term literally means 'open wedding' and refers to the fact that reproduction in these plants was clearly understood unlike the case in lower plants (Cryptogams) in which it was for a long time something of a mystery.

- **phenetic similarity**

same as overall similarity.

- **phenolics**

plant compounds (structurally characterised by an alcohol group on an aromatic ring) that impart a variety of functions to plants, including defence mechanisms and interactions with other organisms. Phenolics can also determine plant properties such as flavour and palatability.

- **phenology**

the branch of science that deals with the influence of climate on periodic biological phenomena, such as the flowering of plants.

- **phenotype**

1. the physical characteristics and appearance of an organism.

2. the outward form, characteristic and appearance of a plant, produced by the interaction of environmental and situational factors upon the traits dictated by the plant's genes.

3. the visible or measurable characteristics of an organism.

4. (Gr. *phainein*, to appear *typos*, image) the characters of an organism due to the interaction of genotype and environment, a group of individuals exhibiting the same phenotypic characters. The detectable expression of the interaction of genotype and environment constituting the visible characters of an organism.

- **phenotypic**

a set of characters arising from reaction to environmental stimulus.

- **pheremone**

a chemical released by insects that attracts members of the same species. Pheremones are usually sexual in nature and are used to lure insects for trapping, monitoring populations or to disrupt mating.

- **pheremone trap**

any of various containers baited with sexual scent, used to lure and entrap garden pests.

- **phloem**

1. the main food-conducting tissue of vascular plants, comprising sieve elements, parenchyma cells, fibres and sclereids.
2. photosynthate conducting tissue of vasular plants.
3. the photosynthate-conducting tissue of plants.
4. (Gr. *phloios*, inner bark) the tissue involved in the transport of carbohydrates and food materials in a vascular plant, being composed of sieve elements, parenchyma cells and sometimes also of fibres and sclereids.

- **phosphate**

pO_3^{4-} a mineral that often participates in permineralisation.

- **phosphorous**

an element used by plants to promote root and tuber growth and the production of flowers and seeds. In a commercial fertiliser, its proportion is represented by the second of three numbers, 5-10-5 fertiliser contains twice as much phosphorous as potassium or nitrogen.

- **photonasty**

of a movement in response to variation in light intensity.

- **photoperiod**

the relative lengths of alternating periods of darkness and light affecting the growth and maturity of an organism, as flowering or the dropping of leaves of deciduous plants.

- **photoperiodism**

the ability to respond to the lengths of alternating periods of light and darkness.

- **photosynthesis**

1. the production of sugars from water and carbon dioxide using energy trapped from sunlight with the aid of chlorophyll.
2. this is the chemical reaction by which carbon dioxide and water combine producing oxygen and the sugars that the plants need for growth. This is accomplished by light acting on light-sensitive pigments called *chlorophylls*.
3. the process whereby sunlight is converted into energy by chlorophyll.

- **phototaxis**

1. movement of a whole organism or freely m otile part in response to variation in light intensity and direction.
2. the orientation of a plant or plant parts with reference to the direction of the source of light.

- **phototropism**
  bending, growth or movement of a part towards or away from light.

- **phragmoplast**
  microfibrils parallel to the spindle axis at telophase across which a cell plate is deposited in cell division

- **phycobilin**
  1. class of pigment found only in algae, comprising red phycoerythrin and blue-green phycocyanin.
  2. biliprotein pigments of Cyanobacteria and Rhodophyta.

- **phycobilisome**
  organelle on surface of thylakoids in which biliprotein pigments are present in Cyanobacteria and Rhodophyta.

- **phycocyanin**
  blue water-soluble biliprotein pigment of Cyanobacteria and Rhodophyta.

- **phycoerythrin**
  red water-soluble biliprotein pigment of cyanobacteria and Rhodophyta.

- **phycoplast**
  assemblage of microtubules perpendicular to the spindle and at the equator of the cell at telophase in cell division.

- **phyllary**
  one of the bracts forming the involucre of flowers of the Compositae.

- **phylloclade**
  a flattened stem or branch that assumes the functions of foliage, as the broad, succulent stems of the Cactaceae (Cactus family).

- **phyllode**
  1. an expanded leaf stalk, which functions as and resembles a leaf blade. e.g. asparagus.
  2. a leaf that consists of an enlarged midrib and lacks blades.
  3. (Gr. *phyllon*, leaf; *eidos*, form) a winged petiole with flattened surfaces placed laterally to the stem and functioning as a leaf.

- **phyllodium**
  broad flat leaf produced on Sarracenias instead of pitchers in response to stress or Winter.

- **phyllody**
  the metamorphosis of a plant organ, such as the corolla of a flower, into a green leaf.

- **phyllogenetic**
  concerned with or relating to the development of leaves.

- **phylloid**
  leaf-like.

- **phyllomania**
  the production of leaves in unusual numbers or in unusual places.
- **phyllome**
  a leaf or plant part that is genetically derived from a leaf.
- **phyllomic**
  pertaining to or resembling a phyllome.
- **phyllophore**
  the top of a palm stem.
- **phyllotaxis, phyllotaxy**
  the study of how repeated structures such as leaves are arranged around a stem, tubercles around a cactus body. Reference to specific arrangements (e.g. (3,5) spiral phyllotaxis). Types of phyllotaxes are spiral, multijugate, distichous and whorled of which the last two are special cases of the first two. (Gr. phyllo = leaf + taxis = arrangement).
- **phyllotaxy**
  1. the distribution or arrangement of leaves on a stem also, the genetically determined laws that govern such distribution.
  2. the pattern of leaf initiation at the apical meristem.
  3. (L. *phyllo-* and Gr. *taxis*, arrangement) the arrangement of the leaves on the stem. The three common positions are alternate, opposite and verticillate.
- **phylogenetic character**
  see **character.**
- **phylogenetic systematics**
  same as cladistics, a method of grouping organisms that is basedsynapomorphies or shared derived traits or characters, compare to numerical taxonomy.
- **phylogeny**
  1. the evolutionary history of a species or plant group.
  2. an hypothesis of evolutionary relationships among organisms. The pattern of lineage branching produced by the evolutionary history of the organisms considered
  3. (Gr. *phyle*, tribe E. *genesis*) the racial history or evolutionary development of any plant or animal species.
- **phylum**
  primary taxonomic grouping, division.
- **physiognomy, foliar physiogonomy**
  analysis of leaf features that are sensitive to environment.
- **phytoclast / phytoclasts**
  a general term for plant-derived, more or less resistant-walled, particle occurring in a sediment.

• **phytoliths**
the siliceous secondary cell walls of certain terrestrial plants. Phytoliths are destroyed by hydrofluoric acid so they are never present in routine palynological preparations and phytoliths processing invoives treatment with strong oxidising agents, which destroy pollen. Studied phytoliths generally range from ca. 20 - 50 μm.

• **phytoremediation**
the use of plants to clean up sites that have been contaminated with chemicals or petroleum products.

• **piceous**
in botany, pitch-black, shiny brownish-black.

• **pileate**
having a pileus or cap, like a mushroom.

• **pileolus**
a small pileus.

• **pileus**
1. the expanded umbrella-like cap of a mushroom.
2. (L. *pileus*, cap) umbrella-shaped structure of mushrooms or toadstools.

• **piliferous**
terminationg in a fine, soft hair.

• **pilose**
1. covered with hair, especially with soft or fine hair.
2. (L. *pilosus*, hairy) hairy, pubescence comprised of scattered long, slender, soft hairs.

• **pilum / pilate**
a sexine element, usually standing directly on the nexine, consisting of a rod-like part (columella) and a swollen apical part (caput).

• **pinching**
removing the terminal bud, e.g. to encourage side-shoots or to prevent flower formation.

• **pinching, pinching back or pinch out**
to remove the growing tip of a plant, with finger and thumb, to encourage the production of sideshoots or the formation of flower buds. Pinching out is also used when small side shoots are completely removed. This is done when single stems are desired, especially when training to form the 'trunks' of standard (tree-form) specimens.

• **pinguicula**
a very widespread genus of over 50 species of carnivorous plant with sticky, trapping leaves capable of limited movement, (Butterwort.)

• **pinna**
1. one of the primary divisions or leaflets of a divided leaf or frond, such as in ferns and palms.
2. a primary division of a fern leaf.
3. (L. *pinna*, feather) a leaflet or a primary division of a compound leaf. pl. pinnas or pinnae.

• **pinnae**
plural (primary) pinna. First order of subdivision of a compound leaf or fern frond.

• **pinnate**
1. feather-formed. Said of a compound leaf with the leaflets along both sides of the leaf stalk.
2. resembling a feather in structure with the parts arranged on both sides of an axis. A *compound* leaf in which the leaflets grow in a row on each side of the midrib.
3. compound, with the leaflets arranged on both sides of a common axis.
4. feather shaped, with lateral branches on either side of a central axis.
5. (L. *pinnatus*, feathered) divided in a feathery manner with lateral processes of a compound leaf, having leaflets on each side of an axis or midrib.

• **pinnate-pinnatifid**
pinnate with pinnatifid pinnae.

• **pinnatifid**
1. cut or sectioned to form a feather shape, as the fronds of many ferns.
2. pinnately cut, more than half way to the midvein.

• **pinnation**
the state or condition of being pinnate.

• **pinnatisect**
cut to the midrib in a pinnate manner.

• **pinnulate**
having divided segments that are themselves further divided, as the leaves of some acacias.

• **pinnule**
1. the second order of segments of a compound or dissected leaf leaf or frond.
2. a secondary pinna, one of the ultimate divisions of a doubly divided leaf.
3. a secondary division of a fern leaf.
4. (NL. *pinnula*, fr. L., small feather, small fin) a secondary pinna, one of the ultimate divisions of a bipinnate or twice-pinnate leaf.

• **PIs (Protease Inhibitors)**
proteins that accumulate in the seeds and storage organs

of many plants. PIs protect plants against a variety of insect, bacterial and fungal pathogens.

- **pistil**

1. female organ of flower including stigma, style and ovary.
2. the female part of a flower. A complete pistil consists of three parts *stigma, ovary* and *style* (when present).
3. innermost, bears the ovules, may occur singly or in clusters.
4. (L. *pistillum*, pestle) the unit of female function of a flower, may be comprised of a single carpel or two or more carpels united.
5. the stigma, style and ovary collectively, the female reproductive organs.

- **pistillate**

1. having a pistil. Having pistils but no functional stamens.
2. said of a flower bearing a pistil or pistils but not stamens, may refer also to a plant having only pistillate flowers.

- **pistilline**

relating or belonging to the pistil.

- **pistillode**

a rudimentary or vestigial pistil.

- **pit**

a region where the secondary cell wall is absent, but the primary cell wall is present.

- **pit connection**

discrete len-shaped plug of cytoplasm held between two adjacent cell walls in the Rhodophyta.

- **pitahaya**

edible fruit from several species of cactus (e.g. Hylocereus polyrhizus, H. undulatus, Stenocereus queretaroensis, Acanthocereus sp.) is commonly called *pitahaya, pitajaya, pitaya*. The fruit of all cacti is safe to eat, avoiding spines, glochids and tiny hairs present on fruit of some species, but only a few species have any culinary merit. Fruit of Acantho-cereus species is said to have diuretic properties. Cactus fruit is farmed and exported from several countries including Colombia, Honduras, Israel, Mexico and Nicaragua and is sometimes available in European and USA supermarkets.

- **pitcher**

1. in botany, a specially adapted pitcher-shaped modification of the leaf of certain plants, such as Nepenthes and Sarracenia (Pitcher Plants).
2. a pitfall trap formed from a leaf modified by rolling with the sides fused into a keel.

- **pitcher plant**
an unspecific name applied to plants of the genera Cephalotus, Heliamphora, Nepenthes and Sarracenia.

- **pith**
1. the central parenchymatous tissue in a vascular plant axis.
2. ground tissue in the center of a stem.
3. (AS. *pitha*, pith) the soft, spongy tissue, consisting of cellular tissue, in the centre of certain plant stems.

- **pitted**
a general term for small depressions.

- **placenta**
1. the part of the *carpel* of flowering plants that bears the *ovules*.
2. (L. *placenta*, flat cake) the part of the ovary from which the ovules arise. It generally occupies the whole or a portion of an angle of a cell.

- **placentation**
1. the manner in which the placenta is arranged in the ovary.
2. the formation of ovules on the wall of the ovary.

- **placentiform**
thick-rimmed round body, concave on upper and lower surfaces.

- **planaperturate**
describing a pollen grain with an angular outline, in which the apertures are situated in the middle of the sides when seen in polar view, rather than at the angles. Example: Tilia (Tiliaceae).

- **plane**
a perfectly flat surface.

- **plankton**
drifting organisms.

- **planktonic**
free-floating.

- **plano-convex**
flat on one side and convex on the other.

- **planozygote**
a motile zygote.

- **plant**
1. a living organism, in the Kingdom Plantae, with cellulose cell walls and lacking a nervous system or powers of voluntary motion. Self-supporting plants use chlorophyll to capture energy from sunlight, using it to synthesise food from carbon-dioxide, water and other inorganic substances. Dependent plants (e.g. fungi, parasites, saprophytes) have no chlorophyll and use pre-existing carbon compounds as a food source. A complete plant may have roots, stems and leaves but

in some cases consists of only a single leafy tissue, a series of linked cells or even a single cell. Plants are divided into flowering plants (angiosperms) reproducing through seeds and flowerless plants (gymnosperms) which reproduce by single celled spores.

2. (L. *planta*, plant) any of a kingdom (Plantae) of living beings typically lacking locomotive movement or obvious sensory organs, generally making its own food, possessing cell walls and unlimited growth.

- **plant breeding**

the propagation of selected plants to develop unique qualities such as prettier flowers or disease resistance.

- **plantlet**

1. a new plant which usually grows from a wound in a leaf or stem or is naturally produced on the leaf edges as in some varieties of Kalanchoe.

2. a little plant.

- **plasmodesma (ta)**

cytoplasmic channels lined with plasma membrane that connect the protoplasts of adjacent cells across the cell wall.

- **plasmodesmata**

cytoplasmic strands passing between the protoplasts of adjacent plant cells through pores in the cell walls.

- **plastid**

1. a discrete structure, such as a chloroplast, within a cell often containing water-insoluble coloured pigments such as chlorophyll and associated yellow, range or red pigments.

2. a group of organelles characterised by a double membrane envelope and a complex of internal membranes. Plastids contain DNA and replicate autonomously.

- **plastochron**

the time between the initiation of one leaf and the initiation of the next.

- **platea / plateae**

areas of the inner wall layer separated by a triradiate channel extending between the endogerminals of a Normap-olles pollen grain. Example Pompeckjoidaepollenites.

- **platea luminosa / plateae luminosae**

synonym of groove.

- **platyspermic**

a flattened seed with bilateral symmetry.

- **pleach**
  to join (the branches of vines, shrubs etc.) by braiding or weaving them together.

- **plectostele**
  a protostele dissected into anastomosing plate-like units.

- **plesiomorphy, plesiomorphic trait**
  same as primitive trait.

- **pleurocarpous**
  prostrate (used for mosses).

- **pleurotreme**
  synonym of planaperturate.

- **plica**
  a general term for a fold, applied in palynology to ridge-like folds of the exine in Ephedra (Ephedraceae) and Lusatipollis.

- **plicate**
  1. folded lengthwise like a fan, pleated.
  2. (L. *plicatus*, to fold) folded into plaits, usually lengthwise, arranged in pleats, as a fan.

- **plumose**
  feather-like or feathery or plume-like, with fine hairs on opposite sides, as in the *pappuse* of some Compositae.

- **plumose**
  (L. *plumosus*, feather) with hair-like branches, feathery.

- **plumule**
  the shoot portion of the embryo.

- **pluricolumellate**
  with the columellae arranged in several rows beneath each murus.

- **plurilocular**
  having many small chambers or locules.

- **pneumatocyst**
  gas bladder or float as in the brown alga Macrocystis.

- **PNTS (plants with novel traits)**
  any genetic plant strain with characteristics that show neither familiarity nor substantial equivalent traits to plant species already cultivated in Canada.

- **pocosin**
  (Algonquian) a bog that has formed in a shallow, undrained depression, the surrounding land being somewhat elevated, the vegetation predominantly evergreen shrubs or small trees. Pocosins vary greatly in size.

- **poculiform**
  cup-shaped.

- **pod**
  a dry fruit or seed vessel that bursts open when mature and is more or less elongated and cylindrical or flattened, as of the pea, bean or catalpa.

- **podzolisation**
  the process by which soil is depleted of alkaline material and becomes more acidic.

- **polar axis / polar axes**
  the straight line between the distal and proximal poles of a pollen grain or spore.

- **polar lacuna / polar lacunae**
  a lacuna situated at the pole of a lophate pollen grain.

- **polar view**
  a view of a pollen grain or spore in which the polar axis is directed towards the observer.

- **polarity**
  direction of evolutionary change, evolutionary order, i.e which character state follows another in a character transformation series.

- **pole**
  either of the two extremities of the polar axis.

- **pollard**
  a tree cut back nearly to the trunk in order to form a thick head of spreading branches, which are cut for basket-making and kindling. Poplars and Willows are often pollarded.

- **pollen**
  1. the dust-like particles on the *anthers* of a flower are pollen grains, these contain the male gametes. They must unite with the female gametes before fertile seeds can be produced.
  2. the male gametophyte, includes two cells.
  3. the plant equivalent of sperm produced in abundance by anthers, the male reproductive organs found in most flowering plants. Immature pollen grains called microspores have significant biotechnological implications.
  4. (L. *pollen, pollis,* fine flour) the male or fertilising element of seed plants, consisting of fine yellowish powder formed within the anther of the stamen.
  5. the multinucleate reproductive microgametophyte of seed plants, enclosed in the microspore wall. Fossil pollen consists of the microspore wall, comprised of sporopollenin and generally identified by its apertures and sculpture. Unlike 'spore' the word 'pollen' is a collective noun like the word 'flour' from which the word palynology is derived. Therefore, the plural 'pollens' should not be used.
  6. the microgametophyte of seed plants, developed from the microspore.

- **pollen analysis**
  the study of assemblages of dispersed palynomorphs such as

those isolated from samples of peat.

- **pollen cement**
  synonym of pollenkitt.
- **pollen class / pollen classes**
  an artificial grouping of pollen grains that share a distinctive character or suite of characters. Such classes are useful in identification keys and may be subdivided into more restrictive categories, pollen types and pollen groups. Example tricolpate class.
- **pollen group**
  a pollen morphological category, subsidiary to a pollen type, including a number of pollen grains that show intergrading characters but no distinguishing characters.
- **pollen mother cell**
  synonym of microsporocyte.
- **pollen rain**
  the expected production and dispersal patterns of spores and pollen for any given area.
- **pollen sac**
  one of four cavities in an anther that contain pollen.
- **pollen tube**
  this is the tube that develops from the wall of the pollen grain in seed plants, providing a passage way for the male gametophytes to reach the embryo sac for fertilisation.
- **pollen type**
  a pollen morphological category, subsidiary to a pollen class and including pollen grains which can be distinguished either by one distinct character or by a unique combination of characters.
- **pollenkitt**
  a sticky material, produced by the tapetum, that may hold pollen grains together during dispersal.
- **pollinarium**
  a functional unit in Orchid pollination that consists of two or more *pollinia*, stalk or stipe and viscidium.
- **pollination**
  1. the transfer of pollen from the pollen organ to the ovule, for example in flowering plants fromstamens to the stigmatic surface of the carpel.
  2. the transfer of pollen from the *anther* to the *stigma* of the same or a different flower to ensure fertilisation of the *ovules* to produce seeds.
  3. the process of fertilisation whereby pollen is transferred from the stamen to the stigma.

- **pollination ecology**
  the branch of agriculture or ecology concerned with the distribution of pollen by wind or animals and its efficacy in fertilisation and seed set.

- **pollinium**
  1. a mass of coherent pollen characteristic of orchids and milkweeds.
  2. a mass of pollen grains.
  3. a general term for aggregations of many pollen grains, which form dispersal units.

- **polumbra**
  a darkened triangular or subcircular area centred on the proximal pole. Example Retusotriletes distinctus.

- **polyacantha**
  with many spines.

- **polyad**
  a dispersal unit comprising more than four pollen grains. Example Acacia (Mimosaceae).

- **polyadelphous**
  having the stamens united in three or more bundles.

- **polyancistrus**
  with many hooks.

- **polyandrous**
  having many stamens.

- **polyannulus / polyannuli / polyannulate**
  a structure in which the sexine of the outer aperture has multiple layers each with its own thickening. Example Atlantopollis.

- **polyarch**
  a protostele with many arms of xylem, most common in monocots.

- **polycarpellary**
  composed of several carpels.

- **polycarpic**
  1. a plant that flowers more than once in its lifetime.
  2. producing fruit many times or indefinitely; having a gynoecium forming two or more distinct ovaries or carpels.

- **polycephalus**
  with many heads.

- **polychotomosulcate**
  describing a pollen grain with a many branched sulcus.

- **polycotyledon**
  a plant having more than two cotyledons or seed-leaves. This is usually the case with pines and most conifers.

- **polycotyledony**
  a variation in the number of cotyledons of a plant.

- **polygamodioecious**
basically *dioecious*, but also having some bisexual flowers present in some or all plants.

- **polygamo-dioecious**
polygamous but chiefly dioecious.

- **polygamo-monoecious**
polygamous but chiefly monoecious.

- **polygamous**
1. bearing both unisexual or bisexual flowers on the same plant or on plants of the same species.
2. (Gr. *polys*, much or many *gamos*, marriage) having bisexual, pistillate and staminate flowers on the same individual plant.

- **polyglucan granules**
polymers of glucose somewhat like glycogen in animals.

- **polygynoecial**
having a number of pistils joined together, as in aggregate fruits such as the raspberry.

- **polygynous**
having many styles.

- **polymorphic**
1. occurring in several different forms, as leaves on the same plant that have different shapes.
2. (Gr. *polys*, many *morphe*, shape) having, assuming or occurring in various forms, characters or styles.
3. of very variable shape.

- **polymorphism**
the capability or tendency to have wide variation in form.

- **polymorphous**
having or assuming many forms.

- **polypetalous**
1. having two or more separate petals.
2. (Gr. *polys*, many *petalon*, a petal) with many separate petals.

- **polyphyletic**
of a group of organisms not including the most recent common ancestor, often because this common ancestor is not characteristic of the group or because the group originated from more than one evolutionary line. Polyphyletic taxa are considered 'unnatural' and tend to become targets for revision by taxonomists.

- **polyphyly, polyphyletic group**
a group of organisms with different most recent ancestors.

- **polyplicate**
describing a pollen grain with more than three meridional ridges (plicae) separated by deep grooves. Example Ephedra (Ephedraceae).

- **polyploid**
  with more than two sets (diploid) of the basic chromosome number.

- **polypompholyx**
  a former genus of carnivorous plant from Australia, currently incorporated in Utricularia.

- **polyporate**
  with many pores. Synonym of pantoporate and zonoporate.

- **polysaccharide**
  1. a long branched or unbranched chain of sugars linked together by glycosidic bonds. Polysaccharides are frequently molecules of great size, with molecular weights up to several million and may serve architectural (e.g. cellulose) or food storage (e.g. starch) functions.
  2. a polymer composed of sugars.

- **polysiphonous**
  having many filaments or pericentral cells as in the red alga Polysiphonia.

- **polysporangium**
  1. a sporangium containing many spores.
  2. a sporangium producing many spores, homologous with a tetrasporangium.

- **polystamonous**
  having many stamens, polyandrous.

- **polystelic**
  having more than one stele.

- **polystichous**
  1. with leaves in several rows.
  2. type of thallus constructed of true parenchyma as in some brown algal orders.

- **polystromatic**
  blade of many cell layers.

- **polytomous**
  divided into more that two subordinate parts or branches, pinnatifid.

- **polytomy**
  the state of being divided into more than two parts.

- **pomologist**
  a person who is experienced in pomology, a cultivator of fruit trees.

- **pomology**
  a department of knowledge that deals with fruits, a branch of gardening that deals with the cultivation of fruit trees or fruit-bearing shrubs.

- **pond**
  (form of pound, enclosure) a body of standing water smaller than a lake, often artifically formed.

- **pontoperculum / pontopercula / pontoperculate**
  a type of operculum that is not completely isolated from the remainder of the sexine but linked to it at the ends of the aperture. Example Sanguisorba officinalis (Rosaceae), Passieflora tetandra (Passiefloraceae).

- **poral lacuna**
  a lacuna of a lophate pollen grain surrounding an endoaperture, which communicates with adjacent abporal lacunae via interlacunar gaps. Example Cichorium intybus (Compositae).

- **pore**
  1. a small opening, as in the surface of a leaf.
  2. a general term, applied in palynology to a circular or elliptic aperture with a length/ breadth ratio less than 2.

- **pore canal**
  the space between the ectopore and the endopore.

- **pore membrane**
  the aperture membrane of a pore.

- **pore space**
  the total space within soils that isn't occupied by soil particles.

- **poro-colpate**
  describing a pollen grain with an arrangement of apertures in which colpi alternate with pores round the equator. Example Pardoglossum (Boraginaceae).

- **pororate**
  describing a pollen grain with compound apertures in which both the ectoaperture and the endoaperture are pores and the two are not congruent. Example Myrica gale (Myricaceae).

- **porostiy**
  the degree to which soil mass is permeated with pores or cavities.

- **porous subsoil**
  a subsoil in which air between its particles allows the water to drain through easily.

- **porrect**
  extended horizontally, stretched out, pointing outward and forward.

- **porus / pori / porate**
  synonym of pore.

- **posterior**
  towards the back or towards the main axis.

- **pot feet**
  small supports placed under pots and planters to raise them

off the ground for better drainage and air circulation.

- **potassium**
an element used by plants for growth, often present in the soil but not in the right form to be absorbed by the plant's roots unless humus (decayed organic matter) is present.

- **pot-bound**
said of a plant having an overly extensive root system in a too-small container.

- **p-protein**
a stringy protein within sieve elements that blocks sieve plate pores when the sieve tube is damaged.

- **praemorsous**
bitten, ragged as if chewed.

- **premorse**
appearing to be broken off at the end, as a root or stem.

- **prepollen**
the microspores of certain extinct seed plants characterised by proximal apertures and presumed proximal germination, rather than the distal, equatorial or other typical apertures of seed plant pollen grains.

- **pricking off**
a method of transplanting tiny seedlings. The blade of a pocket knife or plant marker is used to remove each plant from one spot and move it to another.

- **pricking out**
a method of thinning seedlings by cutting them off at soil level so as not to disturb the roots of the other plants.

- **prickle**
1. a small, spine-like point arising from the bark or epidermis of a plant, instead of from the wood.
2. (ME. *prikle, prikel,* fr. OE. *prickle, pricel* a kin to MD. *prikel,* prickle) a sharp pointed emergence arising from the epidermis or bark of a plant.

- **primary cell wall**
a cell wall layer deposited while a cell is growing, typically extensible.

- **primary growth**
growth in length, controlled by the apical meristem.

- **primary meristem**
a meristem that is present in the embryo of a plant, generally responsible for increase in the length of plants.

- **primary meristematic tissue**
a group of cells beneath the apical meristem that has become distinct in appearance from

neighbouring groups of cells, a precursor to one of the tissue systems.

- **primary pinnae**
leaflet of first subdivsion of a fern frond or compound leaf.

- **primary production**
the quantity of new organic matter created by photosynthesis.

- **primexine**
a developmental precursor of at least part of the exine (the sexine/ectexine), formed during the tetrad stage that is composed largely of polysaccharides and therefore lacks resistance to acetolysis.

- **primitive trait**
same as plesiomorphy a character that is present in the common ancestor of a clade. A primitive trait is inferred to be the original character state of that character within the clade under consideration. compare to derived trait.

- **primordium**
a cell or organ in its initial stage of development.

- **prismatic**
prism-shaped. Having several longitudinous angles and intermediate flat faces.

- **pro-**
a prefix indicating a developmental precursor of a feature. Example probaculum.

- **proboscidate**
beaked, similar to corniculate.

- **procambium**
1. the primary meristematic tissue that gives rise to primary xylem and primary phloem, procambia are found in apical as well as intercalary meristems.
2. develops into the vascular tissue system.

- **procarp**
carpogonial branch adjacent to one or more auxiliary cells in Rhodophyta.

- **process**
in botany, this refers to an organic structure arising from the surface of a plant or plant part, such as a spine on a stem.

- **procumbent**
1. lying on the ground but not rooting.
2. trailing or lying flat without taking root.
3. trailing, lying flat along substratum, prostrate
4. (L. *pro*, forward; *cubare*, to lie down) trailing or lying flat but not rooting.

- **procumbent cells**
elongated parallel to the ray.

- **production**
(L. *producere*, to produce) the weight of new organic material formed over a period of time, plus any losses during that time period. Losses may be due to respiration, excretion, secretion, injury, death or grazing.

- **productivity**
1. the present capability of a kind of soil for producing a specified kind of plant.
2. amount of production over a given period of time. Expressed as a rate such as g/m2 per day, kg/ha per year etc.

- **profile**
a vertical section of the soil through all its layers.

- **projectate**
describing a pollen grain in which the apertures are borne on the ends of strongly projecting arms. Example Aquilapollenites.

- **prokaryote**
simple organism without a defined nucleus or internal membranes.

- **prolate**
describing the shape of a pollen grain or spore in which the polar axis is larger than the equatorial diameter.

- **prolate spheroidal**
describing the shape of a pollen grain or spore in which the ratio between the polar axis and the equatorial diameter is 1.00-1.14.

- **proliferous**
1. bearing vegetative offsets. Used to describe flowers or fruits that produce further flowers.
2. (L. *proles, prolis*, offspring *ferre*, to bear) bearing supplementary structures such as buds or flowers, either in an abnormal manner or in a manner that is normal but from adventitious tissue.

- **promoter**
target binding sites on a DNA strand for RNA polymerase or other proteins which control gene expression.

- **prop roots**
roots that help support a plant from above ground.

- **propagating bench**
this is a stationary, shallow box, usually filled with fine sand, but sometimes soil, which is kept moist. Cuttings or slips are inserted into the sand or soil until they form roots.

- **propagating case**
a device used to root cuttings and raise seedlings. It could simply consist of a flowerpot or box cov-

ered with a glass pan or be a specially constructed *frame* with brick or wood sides and a sliding or hinged sash (top).The purpose of this case is to provide a humid atmosphere, which prevents the cuttings from wilting before they've formed roots. To hasten the formation of roots, bottom heat is often provided by hot-water pipes or electric soil-heating cables. Less cuttings will fail, the faster the roots form.

- **propagation**

the various methods by which plants are increased. Following are some different techniques of propagation. Seeds, division or separation, softwood cuttings or slips, grafting or budding, layering etc.

- **propagator**

a structure that provides a humid atmosphere for seedlings, cuttings or other plants being propagated.

- **propagule**

1. a shoot, such as a sucker or runner, which may be used for propagation.
2. multicellular structure for asexual reproduction.

- **propagulum or propagule**

(L. *propages*, layer of a plant) a runner or sucker used in the asexual propagation of plants. pl. propagula or propagules.

- **prophyll**

a rudimentary leaf at the base of a leafy shoot.

- **prophyllum**

a secondary bract on a flower stalk A two-edged initial bract on the flowering part of a plant.

- **proplastids**

specialised for dividing to form new plastid, usually found in meristematic cells.

- **prostrate**

1. lying flat upon the ground.
2. (L. *prostratus*, pp. of *prosternere*, to lay flat) growing on the ground, trailing.
3. lying horizontally.

- **protective layer**

the layer of cells that produce suberin to seal the petiole before abscission.

- **protein**

a large biological macromolecule made from linear chains of amino acids (polypeptides) linked together by amide bonds. Proteins can have structural or functional (e.g. enzymes) roles in cells and large proteins may be formed from multiple polypeptide subunits. The structure of each protein or polypeptide subunit is specified by a single gene.

- **proteomics**
  a study of an organism's proteins, including the molecular structure of the protein. Protein structure often determines the roles that proteins play in plant physiology.

- **proteranthous**
  having flowers that appear before the leaves of a deciduous plant, as in Jacaranda mimosifolia and many species of Prunus (which contains the Peach, Almond, Cherry, Apricot and Plum species.)

- **prothallia**
  these are formed when *spores* come into contact with moisture. Prothallia are thin, green, heart-shaped structures that bear many root hairs which attach to the soil and extract nourishment. Underneath the prothallia, small male and female organs are produced. The male organ is small with two cilia (hairs) which propel it through a film of moisture to the female, where fertilisation takes place. Once that happens, a Fern, which has a single Fern-shaped leaf, is produced and eventually grows into a complete plant.

- **prothallus**
  1. the growth, typical of pteridophytes, following a germination of a spore which for the purpose of sexual reproduction produces the antheridia and archegonia.
  2. gametophyte of lower vascular plants.

- **protoderm**
  develops into the dermal tissue system.

- **protogynous**
  having the stigma mature before the anthers.

- **protogyny**
  1. the condition of having the stigma mature before the anthers.
  2. (Gr. *proteros*, fore *gyne*, female) development of the female organs before the appearance of the corresponding male products - thus inhibiting self-fertilisation.

- **protonema /protonemata**
  the filamentous (moss) or thallose (fern) gametophyte stage of non-seed plants.

- **protoplast**
  the plasma membrane, nucleus and cytoplasm of the cell. A plant cell from which the rigid cell wall has been removed.

- **protosaccus / protosacci / protosaccate**
a saccus which is completely filled with an alveolar structure. This feature is used in the description of pollen in the Perm - Trias. Example Luekisporites virkkiae.

- **protostele**
1. a type of stele with a solid core of primary xylem.
2. stele having a solid column of vascular tissue with xylem centrally located.
3. a single central vascular bundle.

- **protoxylem**
the first primary xylem to differentiate and mature, usually before and during elongation of the axis; protoxylem cells are generally smaller in diameter than metaxylem.

- **protuberant**
bulging outward.

- **proximal**
towards the base of something near to where it is attached.

- **proximal face**
that part of a palynomorph which faces towards the centre of the tetrad, between equator and proximal pole.

- **proximal pole**
the centre of the proximal face.

- **proximocavate**
an exine in which the exoexine is detached or partly detached, from the intexine only on the proximal face.

- **pruinose**
1. having a frosted appearance with a white powdery coating.
2. having a frosty look due to a *bloom* or powder on the surface, said of certain plant surfaces, such as the undersides of leaves or the surface of blueberries or grapes.
3. frosted with white powder.

- **pruning**
trimming branches or parts of trees and shrubs, in order to strengthen those that remain or to bring the tree or plant into a desired shape. Root pruning is also practiced with a spade or other tool, in order to control size, promote fruitfulness or secure a growth of fibrous roots near the stem prior to transplanting.

- **pseudoaperture / pseudoaperturate**
a thinning of the exine which, although superficially resembling an aperture, is not associated with a thickening of the intine and is presumed not to function as an exitus.

• **pseudobulb**
a fleshy enlargement resembling a bulb at the base of the stem, as in many epiphytic orchids. The variously thickened portions of an Orchid stem.

• **pseudocarp**
a fruit, such as an apple or strawberry, which consists largely of tissue other than the ripened ovary.

• **pseudocehalium**
1. a growth of long, hair-like spines on one side of the upper stem of a columnar cactus, such as Cephalium.
2. development of wool and bristles on a mature flowering stem which continues growth. See **cephalium**.

• **pseudocolpus / pseudocolpi / pseudocolpate**
a colpus-like pseudoaperture. Example Myosotis (Boraginaceae.)

• **pseudolamina**
(Gr. *pseudes*, false L. *lamina*, plate) the extended apical portion of a phyllode.

• **pseudoparenchyma**
tissue resembling parenchyma with large, thin-walled cells, but developmentally filamentous.

• **pseudopore / pseudopores / pseudoporate**
a pore-like pseudoaperture.

• **pseudosaccus / pseudosacci / pseudosaccate**
an extensive, saccus-like separation in the wall of a spore resembling a saccus, but lacking the characteristic alveolate infrastructure. Example Grandispora spinosa.

• **pseudospermium**
any one-seeded fruit that is whole at maturity, as in the sunflower.

• **pseudovivipary**
(Gr. *pseudes*, false L. *vivus*, alive *parere*, to produce) a condition where vegetative propagules replace some or all of the normal sexual flowers in the inflorescence.

• **psilate**
describing a pollen or spore with a smooth surface.

• **psilolophate**
describing a lophate pollen grain which lacks spines. See **echinolophate**, **lophate**.

• **pterate**
winged, eg dipterate - 2-winged, tripterate - 3-winged.

• **pteridologist**
a person who is versed in the study of ferns.

- **pteridology**
  the science or study of ferns.

- **pteridophyte**
  1. the general name for the ferns and related genera.
  2. a fern or fern ally such as plants of the group Equisetum.
  3. (Gr. *pteris*, fern *phyton*, plant) a major division of the plant kingdom, having clear alternation of generations with a dominant vascular sporophyte initially dependent upon the gametophyte which is very reduced.

- **pterocarpous**
  having winged fruit.

- **ptychotreme**
  with apertures situated in invaginations of the outline, when seen in polar view.

- **puberulent**
  1. covered with fine, short down.
  2. (L. *pubes*, adult) covered with fine, short and nearly imperceptable down, minutely pubescent, the hairs soft, straight, erect, but scarcely visible to the unaided eye.

- **puberulous**
  somewhat pubescent.

- **pubescence**
  1. a covering of down or hair, as on the surface of a leaf.
  2. the state of having such a covering.
  3. covered with short, soft downy hairs (like a peach).
  4. fuzzy, covered with down or hairs, as the leaves of the African Violet.
  5. (L. *pubescere*, to become mature) a general term for hairiness, covered with soft hair or down.
  6. covered in soft down.

- **pubispina**
  with downy spines.

- **puddled soil**
  thick, massive soil, artificially compacted when wet, a common result of tillage of clayey soil when it is wet.

- **pulchellus**
  small and beautiful.

- **pulp**
  the juicy flesh filling the cavity of a fruit.

- **pulse**
  the edible seeds of cultivated leguminous plants, such as peas, lentils and beans. Any of the various plants that produce pulse.

- **pulse crops**
  grain legumes, such as peas, beans, chickpeas and lentils.

- **pulverant**
  a term applied to ungranulated fertilisers that are largely powdered.

• **pulverulent**
with a dusty coating.

• **pulvinate**
cushion-shaped. Convex or slightly so.

• **punctate**
1. having the surface marked with pits, points or dots.
2. (L. *puntcum*, point) dotted. With depressed dots scattered over the surface.

• **punctulate**
covered with very fine points or dots e.g. the surface of a seed.

• **punctum / puncta / punctate**
a rounded or elongate tectal perforation, less than 1μm in length or diameter.

• **pungent**
1. terminating in a rigid sharp point. With an acrid, biting or sharp smell or taste.
2. tapering evenly to a sharp point.

• **pustular**
having bumps, as though blistered.

• **pustulate hair**
(L. *pustulare*, to blister) hair with an enlarged base.

• **pustule**
a raised bump or area.

• **pychnoxylic wood**
dense wood that contains little parenchyma, typical of *Archeopteris* conifers, compare to manoxylic.

• **pyrenoid**
1. proteinaceous region of chloroplast, sometimes the site of accumulation of carbohydrate reserves.
2. organelle usually within a plastid, usually associated with starch food reserves.

• **pyriform**
pear-shaped. More elongated than turbinate.

• **quadrangularis**
with four angles.

• **quadrangulispinus**
with four angled spines.

• **quadrate**
square or rectangular.

• **quadricostatus**
with four ribs.

• **quadriseriate**
in four ranks or rows, for example, see **branching**.

• **quadrispinus**
with four spines.

• **quasisaccus / quasisacci / quasisaccate**
synonym of protosaccus.

- **quinate**
  having an arrangement of five similar parts together, especially used to describe leaflets.
- **quinine**
  an alkaloid, found in the bark of the Cinchona tree, that is used to treat malaria, a tropical parasitic disease carried by mosquitoes. Quinine is an essential bitter ingredient of certain drinks e.g. Indian tonic water, where it can be recognised by its blue fluorescence under black (ultra-violet) light illumination.
- **quinquefoliolate**
  having five leaflets.
- **raceme**
  1. an unbranched inflorescence of stalked flowers.
  2. (L. *racemus*, bunch) inflorescence having a common axis and stalked flowers in acropetel succession.
  3. a flower cluster composed of individual florets equally spaced along a stem - often cymose.
- **racemose**
  an inflorescence whose growing points continue to add to the inflorescence and in which there are no terminal flowers and the branching is monopodial, as racemes or spikes.
- **rachidian**
  of or pertaining to a rachis.
- **rachilla**
  1. a secondary stem or axis of a spikelet in grasses and sedges.
  2. (Gr. *rhachis*, spine) the zig-zag center upon which the florets are arranged in the spikelet of grasses or in some sedges.
- **rachis**
  1. that portion of an inflorescence, above the *peduncle*, bearing flowers.
  2. the axis of a compound fern blade.
  3. the central prolongation of the stalk (peduncle), the axis through an inflorescence or of a leaf stalk (petiole), the axis through a compound leaf.
- **radial**
  1. peripheral or positioned around the edges. In the *Cactaceae*, a spine positioned on the periphery of an areole.
  2. a general term describing features radiating from a centre. Applied in palynology to the region of a spore beyond the ends of the laesurae.
- **radial symmetry**
  of flowers which can be cut into equal halves along many axes passing through the flower centre.

- **radiate**
  spreading outward in a wheel shape from the center, as leaves around a stem. Bearing ray flowers, as do many members of the Daisy family, Compositae.

- **radiating**
  positioned outwards.

- **radical**
  pertaining to or growing from the root. Emerging from the root.

- **radicate**
  to take root. To cause to take root, to plant deeply and firmly.

- **radicicolous**
  living upon or in roots.

- **radicle**
  an embryonic root. The initial root of an embryo or seedling. The root portion of the embryo.The embryonic root a rootlet.

- **radiculose**
  producing numerous rootlets.

- **radiospermic**
  a round seed, a seed with radial symmetry.

- **radius**
  a general term, used in palynology for a laesura in trilete spores.

- **radix**
  the root of a plant.

- **raised bed**
  a raised area for growing plants, usually enclosed by landscape timbers, in which the growing medium can be accurately determined.

- **rameal sheath**
  leaf sheath on the stem joints, as in *Equisetum*.

- **ramentaceous**
  covered with, consisting of or resembling ramentum.

- **ramentum**
  a thin, brownish, fringed scale that appears on the young shoots or leaves of many ferns.

- **ramet**
  (L. *ramus*, branch) an individual member of a clone.

- **ramicaul**
  the leafy stem of some *sympodial* Orchids, such as those in subtribe Pleurothallidinae distinguished from the rhizome.

- **ramose**
  branched, having many branches.

- **ramuliferous**
  having many small branches.

- **ramulus**
  a small branch or branchlet.

- **random**
  with branches arising from buds without relation to leaves.

- **rapeseed**
  a plant widely grown for its industrial oil in the 1940's. In the 1960's breeding efforts led to the removal of two compounds, erucic acid and glucosinolates, changing the plant to an edible oilseed now called canola.

- **raphe**
  1. he part of the supporting stalk of an ovule that is fused to its outer covering. It is usually shaped like a ridge.
  2. elongate opening in the valves of some pennate diatoms.

- **raphide**
  needle-shaped crystal occurring as one of a closely packed, sheathlike bundle.

- **ray**
  one of the flower stems in an umbel, a flower cluster in which all such stems arise from the same point. A daisy-like or ray flower.

- **ray flower**
  1. one of the outer irregular florets, typically with a single tongue-shaped corolla, on flower heads of some species of *Compositae* (Sunflower family).
  2. one of the petal-like parts of the flowers of many of the Compositae (Daisy family), such as the Sunflower and Aster.

- **ray initial**
  an isodiametric cell in the vascular cambium, produces elements of the ray system.

- **ray parenchyma**
  parenchyma cells of the ray system.

- **ray system**
  cells elongated radially, develop from ray initials.

- **ray tracheids**
  tracheids of the ray system.

- **reaction**
  the degree of acidity or alkalinity of a soil mass expressed in pH.

- **receptacle**
  1. enlarged apex of a flower stalk bearing the parts of a flower or the florets of a composite flower. In ferns, a main stem on which sporangia arise.
  2. the enlarged end of the stem of a flower upon which the reproductive organs of the flower are borne.
  3. point on a leaf where sporangia are attached.
  4. the tip of a floral stem, supports the floral organs.
  5. fertile area where gametangia or sporangia arise as in the brown algal order Fucales.
  6. (L. *recipere*, to receive) the more or less expanded apex of

a floral axis which bears the floral parts.

- **reclinate**
  curved backwards.

- **reclining**
  sprawling on the ground or leaning against something.

- **rectimurate**
  describing a pollen grain or spore with more or less straight muri.

- **rectispinus**
  with straight spines.

- **recurved**
  bent, curved or curled downwards or backwards.

- **reflexed**
  abruptly curved backwards. Applied to petals, sepals, leaf-veins etc.

- **regolith**
  the unconsolidated nature of loose earth materials (soil, sand, loose rock etc.) above the solid rock on the earth's surface.

- **regular**
  1. having the members of each whorl of floral organs arranged in a symmetrical shape, as in the rose and petunia.
  2. composed of evenly arranged parts in a symmetrical fashion.

- **regular flower**
  a radially symmetrical flower in which similar parts are equal in size, form or extent of fusion.

- **regular symmetry**
  regular flowers have radial symmetry; they are divisible into equal halves in 2 or more planes.

- **regulatory sequences**
  DNA sequences which direct transcription to occur in the desired tissues and at the correct stages of plant development.

- **remote**
  (L. *remotus*, to remove) separated from one another separated by intervals or spaces greater than the ordinary.

- **reniform**
  1. an oval with the ends curved around in the same direction like a kidney.
  2. kidney-shaped, a reniform leaf is wider than it is long, with an indentation at the base.
  3. (L. *ren*, kidney *forma*, shape) having the form or shape of a kidney.

- **repandate**
  undulate, but less evenly so.

- **repent**
  (L. *repens*, crawling) said of a stem that is prostrate and rooting at the nodes.

- **replum**
  the framework that divides the various sections of certain see cases, which breaks open when mature and that remains when the sections fall away.

- **residual**
  a property of a substance, such as a pesticide, that allows it to remain in an area for a long time.

- **residual procambium**
  procambium located between mature xylem and phloem.

- **residue**
  the amount of inorganic fertiliser, pesticide or herbicide which remains in plants, soil or groundwater after the growing season in which it was applied. Such residues may damage future crops, render vegetables toxic and pollute groundwater and nearby streams.

- **resin duct**
  a tube lined with cells that secrete resin.

- **resmethrin**
  a synthetic pyrethrin used as an insecticide.

- **respiration**
  a process of releasing energy by oxidising molecules such as glucose to carbon dioxide. Respiration produces useful reactive molecules, such as ATP (adenosine triphosphate) which activates many cellular processes by transfer of its terminal phosphate group.

- **rest period**
  a period of dormancy (induced by withholding water during part of the year) that is needed by some plants in order to bring about bloom the following year.

- **restriction enzymes**
  enzymes that recognise specific DNA sequences and cleave them to produce DNA fragments of varying length. The ability to specifically cleave DNA sequences at known sites underlies many of the advances made over the last several decades in molecular biology.

- **resupinate**
  referring to Orchid flowers which twist 180 degrees during development, with the result that the *labellum* is lowermost.

- **reticulate**
  1. a net-like or criss-cross pattern, often of leaf veins.
  2. (L. *reticulatus*, latticed) like network.
  3. forming a network (often of leaf-veins) multiply divided.

- **reticuloid**
with bacula arranged in a more or less reticulate pattern. Synonym of retipilate.

- **reticulum**
1. any plant structure with a net-like form.
2. a network-like pattern consisting of lumina or other spaces wider than 1μm bordered by elements narrower than the lumina.

- **retinaculum**
(NL. fr. L. that which holds or binds, band, fr. *retinere*, to hold back, retain) the hook-like funicle of a seed of a plant of the family Acanthaceae, a band or band-like structure that holds an organ in place. pl. retinacula.

- **retipilate**
describing a reticulum formed by rows of pila instead of muri. Example Callitriche (Callitrichaceae).

- **retrorse**
1. bent or curved backward or downward.
2. (L. *retrorsum*, backward) having hairs or other processes turned toward the base.

- **retuse**
1. a rounded termination with a small notch forming two small lobes.
2. notched at the tip (of a leaf or flower part).
3. (L. *retuses*, blunted) with a shallow, rounded notch at the apex.
4. a rounded termination with a small notch forming two small lobes, slightly emarginate.

- **retusoid**
describing a spore with prominent contact areas and curvaturae. Example Retusotriletes.

- **revert**
to return to an original state, as when plain green leaves are produced on a variegated plant.

- **revolute**
1. rolled or curled backward or downward, as the edges of the leaves of some plants.
2. (L. *revolvere*, to roll back) said of margins that are rolled backward (toward the abaxial side).

- **RFLPS (restriction fragment length polymorphisms)**
variations or polymorphisms, within specific regions of genomes that are detected by restriction enzymes. Restriction enzymes produce DNA fragments of varying length based on genetic differences at the nucleotide level.

- **rheotropism**
bending, growth or movement of a part under the influence of a water current.

- **rhipidium**
  a fan-shaped flower cluster.

- **rhizanthous**
  appearing to flower from the root of the plant.

- **rhizocarpous**
  having the stems and foliage annual but the underground parts perennial, as in *Iris* and *Gladiolus*.

- **rhizoid**
  1. a filamentous root-like structure on a pteridophyte (fern) prothallus.
  2. a plant structure that resembles a root in function and overall appearance, but not in its internal structure.
  3. hair-like filamentous anchorage/absorbing organ.
  4. a hair-like absorptive organ on gametophytes and rarely on sporophytes.
  5. simple anchoring organ.

- **rhizomatous tuber**
  same as a corm.

- **rhizome**
  1. an underground stem from which roots, aerial stems or leaves can arise directly.
  2. a (usually) underground stem that is horizontally oriented; rhizomes may appear like roots, but have a definite node and internode architecture.
  3. rhizomes are thickened, branching, creeping storage stems. They are solid like corms and tubers, but don't have tunics. Although most rhizomes grow laterally just along or slightly below the soil's surface, some grow several inches deep. Roots grow from the underside of the rhizome and during the growing season new plants sprout from buds along the top. Rhizomes can be increased by cutting the parent into sections, each containing at least one eye. Cannas, calla lilies, corydalises and lilies-of-the valley are examples of rhizomes. Some varieties of iris grow from rhizomes. Others are true bulbs, while others form fibrous roots like most perennials.
  4. an underground stem.
  5. (Gr. *rhiza*, a root) a subterranean horizontal root-like stem sending out leaves and shoots from its upper surface and roots from its lower surface.
  6. an underground plant structure from which both roots and shoots emerge directly.

- **rhizosphere**
  the area around plant roots that is the site of many key interactions between microbes and the plant. It is also the site where nutrients are transformed for take-up by the plant.

- **rhomboid**
diamond-shaped.

- **rhomboidal tetrad**
a uniplanar tetrad in which the proximal faces of two individual members are in direct contact and the remaining two are separated, giving a rhomboidal outline to the tetrad. Example Epipactis palustris (Orchidaceae).

- **rhombus**
(Gr. *rhombos*, object that can be turned) an outline like a rhomboid, a parallelogram with equal sides, having two oblique angles and two acute angles.

- **rib**
1. a leaf vein. A ridge of plant-tissue thicker than its surroundings. A raised surface running vertically, sometimes spiraling and in the *Cactaceae* bearing a row of areoles along its summit. Morphologically, a rib is composed of a continuous line of fused tubercles, which in some cases may be discerned as bulges in the structure.
2. a ridge (usually vertical) on a cactus stem that is formed from the skin. Stem succulents (those plants that retain large amounts of moisture in their swollen stems) are often ribbed, allowing them to expand or contract sideways as their water stocks are restored or depleted. One of the main veins of a leaf or other plant organ.
3. a leaf-vein, a ridge of plant-tissue thicker than its surrounding.

- **ribosome**
organelle responsible for protein synthesis.

- **ricin**
an extremely toxic protein found in the seed of the castor oil plant *Ricinus communis*, a member of the *Euphorbiaceae*, grown commercially for the oil content of the seed and sometimes grown as an ornamental garden annual. Less than a milligram of ricin injected into the blood stream can be fatal.

- **rimula / rimulate**
the sub-equatorial aperture that encircles pollen grains of the Classopolles group. Example Corollina (Classopollis).

- **ring porous**
large vessels present only in inner portion of an annual ring.

- **ringent**
of a monopetalous corolla where the lip of which is unequally divided. Appearing in the form of a gaping animal.

- **riparian**
of plants growing along the side or on the bottom of a water-

course or in other aquatic areas. Living or growing along the banks of a river or other waterway.

- **rivulose**

marked with irregular, crooked or wavy lines.

- **RNA**

ribonucleic acid- an extremely long single-stranded linear polymer, made (transcribed) as a secondary copy of the gene sequences in DNA and also found as the primary genetic material in some viruses e.g. tobacco mosaic virus. RNA is found in cells as three basic types:

- mRNA- copies of gene sequences to be translated into proteins.
- tRNA- transfer RNA specifying and collecting one of 20 individual amino acids for insertion into proteins.
- rRNA- ribosomal RNA associated with proteins in ribosomes, the sub-cellular mechanisms responsible for building proteins from the mRNA sequence.

- **rockery**

a mound made of rocks that have the spaces in between them filled with soil. It is usually used for planting small, choice plants that need quick drainage, such as ferns or alpines.

- **rogue**

a plant that falls short of a standard of horticultural quality required by nurserymen, gardeners etc. To uproot or destroy, as plants which don't conform to a desired standard.

- **root**

1. a structure usually providing a plant with anchorage and a means of absorbing water and nutrients from the substrate. Many *Cactaceae* have a widespread shallow fibrous root system, in addition to any taproot or tuber. This is an adaptation for scavenging water before it evaporates. See **aerial root.**
2. a plant organ that functions in anchorage and absorption in seed plants derived from a bipolar embryo.
3. the part of a plant, normally underground, that absorbs nutrients and anchors the plant into the ground.
4. the portion of a plant axis produced by the root apical meristem.
5. (AS. *wyrt*, root) the part of a plant, usually below the ground, that holds the plant in position, draws water and nutrients from the soil, stores food and is typically non-green.

6. a plant structure usually providing a plant with anchorage and the ability to absorb water and nutrients from the soil.

- **root apical meristem**
  a meristem located at the apex of a root.
- **root ball**
  the roots and soil or soil mix that they are growing in when lifted from a container or the open ground.
- **root cap**
  a thimble-shaped mass of cells that covers the root apical meristem.
- **root cuttings**
  some plants are easily increased by root cuttings. Root cuttings are made by cutting off pieces of root, each about 3 inches long and planting them in a frame, greenhouse or outside.
- **root grafting**
  the process of grafting scions (shoots) directly on a small part of the root of some appropriate stock, the grafted root being then potted.
- **root hair**
  a trichome on the root epidermis made from an extension of an epidermal cell which assists with absorption of nutrients. Specialised unicellular trichome found in roots.
- **root nodules**
  structures that develop on the roots of plants that form symbiotic associations with nitrogen-fixing bacteria.
- **root pressure**
  pressure in the roots which, when the shoot is cut off, will cause liquid to secrete from the root stump, the mechanisms and tissues involved in this process are not clearly understood.
- **root pruning**
  cutting the roots of large plants, mainly trees and shrubs, to force more vigorous growth or to prepare the plant for transplantation.
- **root tuber**
  swollen food-storing roots.
- **root zone**
  the part of the soil in which plant roots grow.
- **rooting compound**
  a powdery substance into which fresh cuttings are dipped before inserting in soil or medium, containing hormones to encourage root growth.
- **rootlet**
  a radice, a little root or small branch of a root.

- **rootstock**
  a robust plant which provides the root system in *grafting*. Also known as a stock. *Scions* and buds are grafted and budded to a rootstock or stock. The elongated and often thick rhizomes of certain perennial herbaceous plants such as the Iris, Aspidistra and Solomon's Seal.

- **rootstock**
  same as a rhizome.

- **rosarian**
  a cultivator of roses, a rose-grower, a rose-fancier.

- **rose form double**
  describes a flower with imbricated petals (laid like roof tiles) showing stamens in a concave center when open.

- **roseate**
  (L. *roseus*, rosy) rose-coloured, rosy.

- **rosette**
  1. a tight cluster of leaves that are arranged around a central apex and grow outward to form a radially symmetrical pattern.
  2. a group of leaves radiating from about the same point, often at ground level at the base of a very short stem or at the tip of longer stems.
  3. (Fr., Dim. of *rose*) a group of organs, such as leaves, clustered and crowned around a common point of attachment.
  4. a circular growth-pattern with leaves growing radially from a central point; rose-shaped.

- **rostellate**
  beaked, tapering in a long, hard point.

- **rostellum**
  part of the median stigma lobe of Orchid flowers.

- **rostrate**
  1. with a beak-like projection.
  2. Beaked, tapering in a long, hard point.

- **rosulate**
  forming a rosette.

- **rotate**
  1. radiating horizontally like the spokes of a wheel.
  2. (L. *rota*, wheel) shaped like a wheel, radially spreading in one plane.

- **rotation of crops**
  a recurring series of different crops grown on the same ground, the order of recurrence in cropping. It's found that the same kind of crop can't be grown on the same soil several years in a row with good results. Therefore, a different kind of crop is made to

follow the one before it in repeated series. (Different soils and climates require different systems of rotation.)

- **rotenone**
a broad-spectrum botanical insecticide derived from several tropical plants. One of the stronger and longer lasting botanicals. Also fairly toxic, especially to fish.

- **rotundate**
rounded, imperfectly orbicular, in the form of or terminated by an arc.

- **row cover**
any of various types of semi-transparent materials that are draped or supported by hoops over vegetable crops. They increase growth, provide frost protection and exclude pests.

- **rubber**
(caoutchouc) an organic (polyterpene) component of the latex produced within the bark of large trees in the family *Euphorbiaceae* (spurges) notably the Pará rubber tree (*Hevea brasiliensis*) native to the Amazonian rain forest in South America. Seeds from the rubber tree were sent in 1876 to Kew gardens, London, England by Henry Wickham, a local planter. Following germination of the seeds in the tropical herbarium at Kew, seedlings were exported to Ceylon (Sri Lanka) and a year later sent from Ceylon to Singapore where the method of tapping the tree for latex was developed. This avoided felling the tree to obtain latex. The rubber tree is grown commercially in Malaysia, Thailand, Indonesia and Sri Lanka but only to a very limited extent in South America. The latex from the rubber tree is an aqueous emulsion, also containing various proteins and is generally coagulated by heat to produce natural rubber. Most of this is vulcanised by heating with sulphur to cross-link the polymer chains and harden the product.

- **rufous**
with a reddish colouring.

- **ruga / rugae**
synonym of colpus.

- **rugose**
referring to a rough or wrinkled surface of a plant part.

- **rugulate**
describing a type of ornamentation consisting of elongated sexine elements more than 1μm long, arranged in an irregular pattern that is intermediate between striate and reticulate. Examples

Sedum (Crassulaceae), Ulmus (Ulmaceae), Camptotriletes corrugatus.

- **rugulose**
  same as rugose.

- **runcinate**
  1. (L. *runcina*, a plane) pinnatified, with the lobes convex before and straight behind, pointing backward, like the teeth of a saw, as in the dandelion leaf.
  2. in the form of overlapping arrowheads, the backwards-facing barbs usually curved, eg a dandelion.

- **runner**
  a horizontally spreading stem that runs above ground and roots at the nodes to form new plants. The stem eventually dies leaving many well-rooted plants surrounding the original one. An example is the Strawberry. Often confused with *stolon*.

- **runner**
  (AS. *rinnan*, to run) a specialised stolon consisting of a prostrate stem rooting at the node and forming a new plant which eventually becomes detached from the parent plant as in a strawberry plant.

- **rupate**
  synonym of loxoaperturate.

- **rynia**
  a slow-acting botanical insecticide derived from the stems of tropical rynia. It is fairly long-lasting (up to 2 weeks) and toxic in concentrated forms. Often combined with other botanicals.

- **sabadilla**
  a broad-spectrum, short residual (2 days in sunlight) botanical insecticide derived from a South American lily. It can be very toxic to humans in purified form.

- **sabulous**
  sandy, gritty.

- **saccate**
  1. furnished with or having the form of a bag, pouch or sac, as a saccate petal.
  2. (L. *saccus*, bag) having the form of a sac, pouchlike.

- **saccate pollen**
  pollen with a saccus or sacci, characteristic of many conifers.

- **sacciform**
  resembling a pouch.

- **saccoid**
  a saccus-like expansion of the exine with a complex infrastructure build up by a three dimensional network of sexine elements, extending to and fused with the nexine.

• **saccus**

a sac formed by an expansion of the exine of a pollen grain and at least partly filled with an alveolate infrastructure.

• **saccus/sacci**

a winglike or bladderlike extension on a pollen grain.

• **sagittate**

1. shaped like an arrow head with two lobes pointing backwards.

2. (L. *sagitta*, arrow) shaped like the head of an arrow with the basal lobes pointing downward.

• **sagittate (adj)**

1. in the shape of an arrow head. In the form of two pointed lobes directed bacwards.

2. derived from *sagitta*, an arrow, means the leaves are shaped like arrowheads.

• **saline**

salty.

• **salsuginous**

growing naturally in soils with a high salt content, halophytic.

• **saltation**

the movement of soil and mineral particles by intermittent leaps from the ground when particles are being moved by wind or water.

• **salverform**

1. noting a corolla with the outer edge spreading out flat, as in the primrose and phlox, hypocrateriform.

2. (L. *salvus*, safe *forma*, shape, figure, image) said of a corolla in which the tube is essentially cylindrical, the lobes abruptly spreading; a gamopetalous corolla.

• **samara**

1. a winged fruit containing a single seed. (Latin *samara* = elm seed).

2. a dry, usually one-seeded fruit provided with a wing, as in the ash, elm or birch. Also called key and key fruit.

• **samariform**

having the form of a samara.

• **saprophyte**

an organism deriving its nourishment from decaying or non-living organic material. A plant which obtains all its nutrients from decaying matter. This may be an animal or vegetation. Saprophpytic flowering plants usually have no green colour in their leaves, which are usually reduced to scales. The roots of these plants are covered by a cobweb-like growth called fungus mycelium. This isn't harmful to the plant, but helps it obtain nutrients from the decaying matter. There are a few flowering plants that are saprophytic, Monotropa and Ericceae are two. There are many saprophytic bacteria and fungi that obtain their nutrients from decaying matter and help the decaying process. The cultivated Mushroom is an example of a saprophyte as it derives its nourishment from decaying manure. A few saprophytic fungi are the causes of disease in plants.

- **saprophytic**
  of a free-living organism obtaining all nutrients from non-living organic sources.

- **SAR (systemic acquired resistance)**
  a long-lasting and broad host-range defense mechanism in plants that can be brought about by previously treating a plant or parts of it, with microbes that induce cell death.

- **sarcocarp**
  the fleshy part of certain fruits, between the epicarp and the endocar, mesocarp, the part of fleshy fruits usually eaten, as in the peach, plum etc.

- **sarcody**
  a hyperplastic symptom in which swellings occur above and below portions of organs that are tightly encircled, as a stem might be 'choked' by a tight tie or binding or by a twining vine.

- **sarcotesta**
  the usually parenchymatous outer layer of the integument in some seeds.

- **sarment**
  a runner, a slender running stem, as that of the strawberry.

- **sarmentose**
  1. having long sarments or runners having the form or character of a runner.
  2. (L. *sarmentosus*, fr. sarmentum, twig; plus, *-osus*, *-ose*) producing slender prostrate branches or runners.

- **sarracenia**
  a genus of 8 species of pitcher plant distributed throughout the South and Eastern coastal regions of the US and North into East and Central Canada. Trumpet pitcher.

- **sartage**
  the clearing of woodland for agricultural purposes, as by setting fire to the trees.

- **satellite DNA**
  short sequences in eucaryotic DNA that are present in large numbers of repetitive copies, differ in base composition and therefore density from most of the DNA and are generally not transcribed. Often associated with the centromere.

- **satellite virus**
  of a small virus associated with tobacco necrosis virus (TNV), which is dependent upon the TNV genome for its own replication. Of a defective virus associated functionally with another virus which it depends on for replication.

- **savanna**
  a plant community dominated by grasses with occasional scattered trees.

- **saxicolous**
  living on rock.

- **saxifragous**
  growing in the crevices of rocks and eventually splitting the rock.

- **saxitoxin**
  neurotoxin produced by dinoflagellate Gonyaulax.

- **scab**
  a hyperplastic symptom of disease characterised by rough, crusty lesions formed by excessive cork production. A disease in which such lesions form.

- **scaberulous**
  minutely scabrous or rough-textured.

- **scabrate**
  describing elements of ornamentation, of any shape, smaller than 1μm in all directions. Examples Quercus (Fagaceae), Artemisia (Compositae).

- **scabrid**
  slightly rough to the touch, as a scabrid leaf.

- **scabrous**
  1. of a leaf, roughened by tubercles, hairs or other structures on its surface.
  2. rough, with rough little dots or scales.
  3. (L. *scabrosus*, rough) with small points or knobs, like a file, scaly, scabby, rough.

- **scalariform**
  contains several elongated perforations such that the remaining cell wall resembles the rungs of a ladder.

- **scale**
  1. a small leaf like structure. A narrow, triangular or spine-like lower perianth segment often found on the ovary or flower tube surface.
  2. the small, initial (often undeveloped), modified leaf/leaves, that forms the protective covering of the leaf buds of deciduous trees, the *involucre* of the Compositae (Daisy family), the bracts of the catkin, the *imbricated* (overlapping) and thickened leaves which constitute a bulb and also applied in the Coniferae (Conifer or Ever-

green family) to the leaves or bracts of the cone.

- **scale insect**
a small flat insect pest, with a waxy dorsal shield, that adheres to the plant epidermis and feeds from the plant sap.

- **scaly**
composed of scales lying over one another, as a scaly bulb having scales scattered over it, as a scaly stem.

- **scandent**
(L. *scandent-, scandens*, pres. part. of *scandere*, to climb - more at scan) climbing. A climbing plant of a creeping or scandent nature.

- **scape**
1. a long flower stalk arising directly from a root. Adj. *scapose*- borne on scapes.
2. a leafless *peduncle*, commonly, the leafless stalk of an entire inflorescence.
3. (L. *scapus*, the shaft of a pillar, the stalk of a plant) a stem growing from the crown of the root, bearing the blossom without leaves.
4. a long flower stalk arising directly from the root.

- **scapose**
1. scape-bearing; scapigerous, consisting of a scape.
2. of a plant whose flowers are borne on scapes.

- **scar**
a mark on a stem or branch seen after a leaf drops or on a seed after it separates form its stalk.

- **scarification**
1. a mechanical seed treatment which helps to break down a hard or impervious seed coat as an aid to germination.
2. having a dry membranous appearance.
3. dry and papery, often said of bracts.
4. (LL. *scariousus*, rough) tough, thin, dry and semitransparent.

- **schizocarp**
1. a dry, compound fruit, which at maturity, splits into two or more, one-seeded *indehiscent carpels*, as in most Umbelliferae.
2. (Gr. *schizein*, to cleave *karpos*, fruit) a dry fruit, as in the maple, that splits at maturity into two or more one seeded carpels which remain closed.

- **scion**
1. the top part of a graft.
2. a shoot or bud of one *species* that is grafted to the *stock* of another plant. A detached living shoot of a plant.

- **sclereid**

1. a short, irregular sclerenchyma cell with pits; sclereids function as tissue support.
2. a cell that has a thick, typically lignified secondary wall and that typically lacks a protoplast when mature, the 'stone cell' of a pear.
3. variable in shape, but not like fibres.

- **sclerenchyma**

1. tissue composed of cells with walls thickened with lignin, sclerenchyma tissue functions primarily in strengthening and support.
2. in higher plants, the protective and supportive tissue made up of cells with *lignified* and thickened walls.
3. tissues composed of cells with thick, secondary cell wall that are usually lignified.
4. (Gr. *skleros*, hard; *enchyma*, an infusion) tissue of uniformly thick-walled, dead cells in the stem whose principal function is mechanical. The cells are usually grouped into fibres.

- **sclerine**

a term encompassing both exine and perine that can be used whether a perine is present or not, sporoderm excluding the intine.

- **sclerophyll - (greek = hard-leaved)**

a woody plant with small, leathery, evergreen leaves that retain water and typically found in hot, dry climates such as Australia.

- **sclerotesta**

the middle, fibrous layer of the integument in some seeds.

- **scolecodonts**

the scleroprotein teeth, jaws and associated features of polychaet annelid worms. Size from 100 - 4000 μm. Present in marine sediments from the Ordovician (ca. 480 Ma) to the Recent, but used for stratigraphic purposes primarily in the Paleozoic.

- **scorch**

1. dry brown discolouration giving a 'burnt' appearance to leaf margins as a result of infection or exposure to unfavou -rable environmental conditions.
2. an injury to foliage caused by excessive *transpiration* or lack of water.

- **scoria**

crushed volcanic rock.

- **scorpioid**

1. having parts arranged in a *circinate* form, as a scorpoid inflo-

rescence in some of the Boraginaceae.

2. (Gr. *skorpois*, scorpion *eidon*, form) resembling a scorpion, said of a circinnately coiled determinate inflorescence in which the flowers are two-ranked and borne alternately at the right and left.

- **scorpioid cyme**
  a coiled inflorescence in which the flowers arise 2-ranked on alternate sides of the axis, as in *Myosotis scorpioides*.

- **scrobiculus / scrobiculi / scrobiculate**
  synonym of punctum.

- **scrotiform**
  1. shaped like a hollow pouch or sac.
  2. pouch-shaped. Hollow, in the form of a sac.

- **scrub**
  1. a plant community dominated by low shrubs.
  2. a bush, shrub, a stunted tree or shrub or one appearing so.
  3. collectively, bushes, brushwood, underwood, stunted forest.
  4. of inferior breed or stunted growth.

- **sculptine**
  a term encompassing both exine and perine (sclerine) but excludes the nexine and so provides a neutral term for the sculptured layer when there is doubt whether the pattern belongs to the exine or the perine.

- **sculpture**
  the external texture or pollen grains and spores . For example spines or reticulations.

- **sculpturing / sculptured**
  the surface relief or topography, of a pollen grain or spore.

- **sculpural density**
  the estimated number of sculptural elements in an area of 100 $\mu m2$ of the surface of the exine.

- **scutate**
  shaped like a small shield, peltate, as a scutate leaf. Lenticulate, but with an elevated rim.

- **scutellate**
  shield or saucer shaped.

- **scutelliform**
  as patelliform, but oval rather than round.

- **scutiform**
  shield-shaped, peltate.

- **sebaceous**
  having the appearance of fat, as the sebaceous secretions of some plants.

- **sebiferous**
  sebaceous, greasy.

- **secondary cell wall**
  innermost layer of a cell wall deposited after cell enlargement has ceased, often lignified.

- **secondary growth**
  growth in width initiated and maintained by the vascular cambium and cork cambium.

- **secondary meristem**
  a meristem that arise from tissues produced by a primary meristem, generally responsible for increase in thickness of plants.

- **secondary phloem**
  photosynthate-conducting tissue produced by the vascular cambium, inner bark.

- **secondary vascular bundles**
  vascular bundles that do not develop from procambium.

- **secondary xylem**
  xylem produced by the vascular cambium, see also xylem and compare with primary and secondary growth.

- **secretory (glandular) trichome**
  secretes a substance.

- **sectile**
  referring to pollinia in loosely coherent 'packets'.

- **secund**
  to one side, as flowers on an inflorescence.

- **sedge peat**
  the peat that results from decaying masses of several types of mosses.

- **sedimentary rock**
  rock composed of particles deposited from suspension in water, like sandstone, limestone, shale and conglomerates.

- **seed**
  1. a sexually-produced embryo, consisting of an axis (root and stem or hypocotyl and epicotyl) and in *Cactaceae* two cotyledons, with its associated protective and nutritive structures capable of developing into a plant.
  2. a fertilised ovule, megasporangium that contains an embryo enclosed in an integument
  3. the ovules of flowering plants (except those of gymnosperms such as Pines) are contained within an ovary. Pollen grains come in contact with the stigma, germinate and send down long tubes through the style into the ovary. These pollen tubes contain nuclei which fuse with the nuclei of the ovules and fertilisation is effected. The ovules then form seeds. Seeds consist of a seed coat (testa), two seed leaves (cotyledons) or one seed leaf and the embryo plant. The embryo consists of a root and a shoot. At the end of the

seed is a scar (hilum), where the seed was attached to the pod. Near this scar is a microscopic hole (micropyle) where the pollen tube entered to fertilise the ovule and through which the root emerges when germination takes place.
4. a mature ovule.
5. (AS. *saed*, seed) the part of a flowering plant that contains the embryo and will develop into a new plant if sown.
6. the sexually-produced embryo and associate protective and nutritive structures capable of developing into a plant.

- **seed coat**

the outermost layer of a seed, develops from the integuments.

- **seed fern**

extinct seed-bearing plants with fern-like leaves.

- **seed leaf**

a *cotylendon*, which is the beginning leaf or leaves in an embryo.

- **seed- megaspore**

a large, functional megaspore associated with three small, presumably aborted spores present in some fossil lycopsids. Example Cystosporites.

- **seed plants**

a monophyletic clade of plants that reproduces by seeds, megagametophyte is retained on the parent sporophyte and enclosed in an integument, microgametophyte is transferred to the megagmetophyte.

- **seed vessel**

the *pericarp* or wall of the ripened *ovary*, which contains the seeds.

- **seedling guard**

a row cover to protect seeds or new sprouts indoors or out.

- **segment**

individual divisions on a pinnule. The ultimate division or unit of a dissected fern leaf.

- **seismonastic**

a movement response to a shock stimulus. Of the movement of the stamens towards the pistil when touched.

- **self-coloured**

uniform in colour, as self-coloured flowers.

- **self-fertile/self-pollinate**

a plant's ability to fertilise its pistils with its own pollen.

- **self-fertilisation**

the fertilisation of a flower by pollen from the same flower. (Compare to cross-fertilisation.)

- **self-seed**
  a plant releases ripe seeds that sprout into new plants without help from the gardener.
- **self-sterile**
  a plant's inability to fertilise itself, as a self-sterile flower; opposite of self-fertile.
- **selliform**
  saddle-shaped.
- **semi-adherent**
  having the lower half adherent, as a seed, stamen etc.
- **semi-amplexicaul**
  half-*amplexicaul*, embracing half of the stem, as many leaves.
- **semi-aquatic**
  living in or close to water, as semi-aquatic plants that grow between tides or in pools that become dry every so often.
- **semi-double**
  this describes flowers that have one two or more rows of regular, irregular or loose petals and conspicuous stamens.
- **semi-erect**
  synonym of subprolate.
- **semilunate**
  tending towards the form of a crescent moon.
- **seminiferous**
  seed-bearing; producing seed.
- **semitectum / semitectate**
  a partially discontinuous tectum in which the tectal perforations are equal to or wider than the muri and usually larger than 1μm in diameter.
- **semi-terete**
  flat on one side, terete on the other.
- **semitransverse**
  synonym of suboblate.
- **sempervirent**
  always green or fresh, evergreen.
- **sepal**
  1. one of the separate perianth leaves of the calyx.
  2. a whorl of a flower, sepals often resemble reduced leaves and function in the protection of the bud, sepals may be modified to function more like a petal
  3. each segment of the *calyx*. Though usually plain and green, it may sometimes be ornate. The outermost whorl of flower parts.
  4. outermost and most leaf-like, usually encloses the rest of the flower in the bud.
  5. (Fr. *sepale* L. *separatus*, separate *pelatum*, petal) any of the leaf divisions of the calyx. When a calyx consists of but one part, it is said to be monosepalous, when of two parts, it is said to be

disepalous, when of a variable and indefinite number of parts, it is said to be polysepalous and when the parts are more or less united, it is said to be gamosepalous.

• **sepalody**
the transformation of petals or other organs into sepals.

• **separate**
one of the individual size groups of mineral soil particles, sand, silt or clay.

• **separation layer**
the layer of cells that secrete cell wall degrading enzymes forming a weak point where an abscising leaf can drop off.

• **sept- or septi-**
seven, as in septifolius - seven-leaved.

• **septate**
divided by or having a septum or partition.

• **septicidal**
(L. *septum*, division *caedere*, to cut or divide) dividing through middle of ovary septa, dehiscing or breaking open at a natural dividing line.

• **septile**
belonging to or relating to septa.

• **septum**
1. (L. *septum*, a partition) a partition separating two cavities or masses of tissue, as in fruits. pl. septa.
2. a partition or dividing wall, as the septum in a compound ovary or fruit.

• **sequencing**
determination of the nucleotide sequence of the DNA. DNA is composed of 4 nucleotides represented by the 'building blocks' A, C, G and T. The order of A, C, G and T within the DNA strand determines the start and stop, when the gene is expressed and most importantly provides the information specifying the amino acid sequence of the protein to be produced from that gene.

• **sericeous**
silky.

• **serotinal**
appearing or occurring in the later part of the summer.

• **serpentine rock**
a rock consisting of acid magnesium silicate.

• **serrate**
1. leaves and other plant parts that have sharp 'teeth' along their edges pointing toward the tip.

2. (L. *serrare*, to saw) notched on edge like a saw, having sharp notches along the edge pointing toward the apex, as, a serrate leaf. When a serrate leaf has small serratures upon the large ones, it is said to be double serrate, as in the elm. A serrate-ciliate leaf is one having fine hairs, like eyelashes, on the serratures. A serrate-dentate leaf has the serratures toothed.

- **serrated**

1. bearing straight-sided teeth. Serrate- with edges like a saw blade.
2. furnished with straight-sided teeth.

- **serrulate**

finely serrate, denticulate, minutely serrate.

- **sessile**

stalkless, immobile. Leaves and other plant parts that have no stalk, but instead grow directly from the stem of the plana. (L. *sedere*, to sit) sitting directly on base without support, stalk, pedicel or peduncle. Attached or stationary as opposed to free living or motile.

- **seta**

1. a hair-like extension of the leaf, as in homophyllous species of *Selaginella*.
2. stiff hair, projection or bristle.
3. (L. *seta*, bristle) a bristle-like structure.

- **setaceous**

bearing bristles. Having or consisting of bristles or resembling a bristle, setiform.

- **setiferous**

setaceous.

- **setiform**

bristle-like.

- **setigerous**

setaceous.

- **setose**

terminating gradually in a fine sharp bristle.

- **setula**

a short seta, bristle or hair.

- **setulose**

finely setose, covered with small hairs or bristles.

- **sewage sludge**

an organic fertiliser that consists of dried processed sewage.

- **sexfoil**

a leaf with six leaflets, a flower with six perianth segments.

- **sexine**

the outer, sculptured layer of the exine, which lies above the nexine.

- **sexine 1, 2, 3, 4, 5**

a system of sexine stratification in which sexine 1 is the innermost and sexine 5 in the outermost layer of the sexine.

- **sexual**
  of reproduction through the merging of male and female gametes to form a zygote which develops into a new plant.

- **shade cloth**
  a fabric stretched over young plants to protect them from excessive heat and moisture loss. Any of various fabrics used in the summer to lower soil temperatures, accelerate germination of cool-season fall crops, prevent bolting and protect against drying.

- **shade netting**
  mesh material that is stretched over a frame to shield plants from too much sun or heat.

- **shaggy**
  pubescent or downy with long and soft hairs.

- **shallow soil**
  soil that has little room for water storage and that can easily be moved by the weather.

- **shape classes**
  categories of pollen and spore shape based on the relations between polar axis (P) and equatorial diameter (E).

- **sheath**
  a tubular covering surrounding an organ or part of a plant, such as the basal part of a grass leaf surrounding the stem.

- **shelterbelt**
  a row or stand of trees, shrubs or other dense growth planted as a barrier between a garden site and the prevailing wind or direction of storms.

- **shoot**
  (ME. *schoten*, to shoot, dart) a young branch which shoots out from the main stock.

- **shoot apical meristem**
  a meristem located at the apex of a shoot.

- **short shoot**
  shoots that are characterised by short internodes giving the shoot a crowded appearance, compare with long shoot.

- **short-day plant**
  1. a plant that tolerates short daylight hours. A plant which doesn't need long periods of sunlight. Chrysanthemums are examples of short-day plants.
  2. a plant requiring less than 12 hours of daylight in order for flowering to occur.

- **shrub**
  1. a horticultural term for a low bushy plant with multiple

stems normally branching from the base, rather than having a single stem branching higher up as in the case of a tree.
2. a woody plant, usually smaller than a tree, which produces several stems rather than a single trunk from the base.

- **shrubbery**
shrubs collectively. A planting of shrubs, as in a garden.

- **shrubby**
shrub-like, scrubby. Abounding in shrubs.

- **shuck**
an outer covering, such as the husk of corn or the shell of a walnut. To remove the husk, pod or shell from.

- **side dressing**
an application of manure or fertiliser that is placed around the plants as they are growing.

- **sieve area**
field of pores through which translocation occurs as in brown algal order Laminariales

- **sieve cell**
1. a phloem conducting cell type in all vascular plants except angiosperms.
2. a sieve element that lacks perforation plates, characteristic of gymnosperms.

- **sieve element**
a conducting cell in the phloem.

- **sieve plate**
1. area of the wall of a sieve tube element that contains several to many perforations that permit cytoplasmic connections between sieve tube cells.
2. the end wall of a sieve-tube element that is perforated by sieve plate pores.

- **sieve plate pore**
an enlarged plasmodesma that perforates a sieve plate.

- **sieve tube**
1. a long tube of sieve elements (also called sieve tube members) connected by sieve plates.
2. longitudinal file of sieve elements as in brown alga Nereocystis.

- **sieve tube cell**
(in angiosperms) a specialised cell derived from the same parent cell as the closely associated companion cell immediately

adjacent to it. Sieve tube cells are elongated cells with sieve plates. Sieve tube cells form sieve tubes through which photosynthate is transported.

- **sieve-tube member**
  a sieve element with perforation plates, characteristic of angiosperms.

- **sigillate**
  having marks resembling seals; for instance, on roots or stems.

- **silica**
  SiO_2. Ocurring in crystalline (quartz), cryptocrystalline (very finely crystalline; crystals are very, very small) (opal) and non-crystalline (chert) forms. One of the most common minerals in the crust of the earth; an important mineral in the process of silification.

- **silica cells**
  cells in the epidermis of grasses that contain silica deposits.

- **silica gel**
  a fluffy powder used as a barrier to household pests, such as ants. It can also be used outside.

- **silicle**
  a short silique, usually about twice as long as wide.

- **silicoflagellate**
  chrysophycean organism with silicious skeleton.

- **silification**
  a type of fossilisation in which silica (SiO_2) infills intercellular spaces (permineralisation) or replaces the cell walls (petrifaction).

- **silique**
  1. the long, pod-like fruit that is characteristic of plants of the family Cruciferae (Mustard family).
  2. (L. *siliqua*, a pod) the long, narrow pod of plants of the mustard family, Cruciferae, with valves which fall away from a frame bearing the seeds.

- **simple**
  undivided. Contains a single perforation.

- **simple leaf**
  this is a leaf that consists of a single blade, unlike a *compound* leaf, which has two or more.

- **simple pit**
  a pit that is not bordered, may be round or slit-shaped.

- **simple tissue**
  tissue composed of only one cell type.

- **simplibaculate**
  synonym of simplicolumellate.

- **simplicolumellate**
  with a single row of columellae under each murus. Example

Viburnum opulus (Caprifoliaceae).

- **single (of flowers)**

this describes flowers which have a single row of petals. However, this may also be used to describe a flower with few petals even though they may be arranged in more than one row.

- **single grain soil**

a structureless soil in which each particle exists separately, as in a dune.

- **sinu-aperturate**

describing a pollen grain in which the equatorial apertures are situated in the middle of concave sides when seen in polar view.

- **sinuate**

wavy-edged. Having a very wavy edge, such as the leaves of *Pittosporum eugenioides*.
(L. *sinuare*, to bend) having a wavy margin, as some leaves.

- **sinuate-dentate**

having a toothed edge that is also very wavy.

- **sinus**

1. the indentations of a lobed pinna / segment of a fern or of a dicot leaf.
2. an indentation or hollow in an edge, such as between two divisions of a leaf.
3. (L. a bent surface, a curve, a fold or hollow, bosom, a bay) the rounded depression between two consecutive lobes, as of a leaf. pl. sinuses, sinus.

- **siphonostele**

1. a type of stele that consists of a ring of vascular tissue surrounding a pith.
2. a stele having vascular tissue in the form of a hollow cylinder, with a central pith.
3. a cylinder of vascular tissue with a central parenchymatous pith.

- **sister group**

two clades that resulted from the splitting of a single lineage; sister groups share a common ancestor.

- **slacked lime**

lime that is used to raise the pH of soil, but is more caustic than dolomitic soil.

- **slight dimorphism**

fertile and vegetative leaves only slightly different in appearance.

- **slip**

a cutting from a mother plant. The down-slope movement of a mass of soil under wet or saturated conditions.

- **slough**
(AS. *sloh*, a slough) a wet place of deep mud or mire. A sluggish channel, a swamp, bog or marsh, especially one that is part of an inlet or backwater.

- **sobole (sah'buh lee)**
a shoot, stolon or sucker.

- **soboliferous**
producing suckers or lateral shoots from the ground, usually applied to shrubs or small trees.

- **sod**
the upper stratum of grassland, containing the roots of grass and any other herbs that may be growing in it. The sward or turf. A piece of this grassy layer pared or pulled off, a turf, a divot or fail.

- **sodium nitrate**
an inorganic fertiliser that adds nitrogen to the soil.

- **softwood**
wood consisting of soft, immature tissue.

- **soil**
the upper layer of earth that can be plowed, dug or tilled. The earth's loose surface material in which plants grow. Land.

- **soil amendment**
a soil additive. What is required for the best plant productivity.

- **soil analysis**
this is a chemical examination of a soil sample to find out its components in order to suggest amendments to improve the soil.

- **soil classification**
the systematic arrangement of soil into groups or categories according to the proportion of sand, silt or clay that is in the soil.

- **soil climate**
the moisture and temperature conditions existing in the soil.

- **soil pasteurisation**
a sterilisation process of soil that destroys unacceptable organisms without chemically altering the soil.

- **solarisation**
a process of sterilising the soil and killing soil pests by covering the moist planting area with clear plastic and letting it 'bake' for up to 6 months. This process is very effective in hot summer climates.

- **sole**
synonym of foot layer.

- **solenostele**
a siphonostele having phloem both internal and external to the xylem.

- **solitary**

vessels separated by other types of cells.

- **solum**

the upper part of a soil profile, above the parent material in which the processes of soil formation are active. Roots, plants and animal-life characteristics of the soil are mainly confined to the solum.

- **somatotropic**

in botany, promoting growth.

- **sorus**

a fruit dot or cluster of reproductive bodies occurring on lower plants such as ferns, lichens and fungi. A familiar example is found in the clusters of sporangia on the undersides of the fronds of ferns.

- **sorus**

cluster of sporangia of a fern. A group or cluster of reproductive structures.

- **sorus, sori**

a cluster or group of sporangia, most frequently applied to clusters of fern sporangia.

- **sp.**

abbreviation for **species**, used as the second part of a binomial name when the genus is known, but not the species.

- **spadiceous**

bearing flowers on a fleshy or succulent spike enclosed in a leaf-like spathe, as do palms.

- **spadiciform**

resembling a spadix.

- **spadicose**

having flowers on or comprising a spadix.

- **spadix**

the fleshy spike usually enclosed in a spathe. The central columns in the 'flowers' (inflorescences) of Calla Lilies, Jack-in-the-Pulpits etc.

- **spathaceous**

having a spathe or having the nature of a spathe.

- **spathe**

1. one or sometimes two, large *bracts*, which grow near the base or surround a flower cluster or individual bud. Examples are the Calla Lily (Zantedeschia), Jack-in-the-Pulpit and Anthurium.

2. (Gr. *spathe*, flat blade) a large leaflike part or pair of such enclosing a flower cluster (especially a spadix).

- **spathulate**

spoon-shaped.

- **spatulate**

1. long and thin broadening towards the tip, like a spatula.

2. spatula-shaped, oblong or rounded with a long, narrow base, as a spatulate leaf or petal. 3. shaped like a spatula or spoon, gradually widening distally and with a rounded tip, as some leaves.

- **special wall**
  pollen mother-cell wall.

- **specialist natural enemy**
  a predator which attacks only one (or a few) species. Specialist natural enemies are sometimes used as biological controls.

- **speciation**
  a process of differentiation of a single species into one or more distinct species. This can occur through geographic isolation, where the species no longer interbreed and their characteristics slowly change under different environmental pressures.

- **species**
  1. a population of individuals recognisable by morphological characteristics and unified genetic relationship, sometimes capable of subdivision into subspecies or varieties which are sufficiently closely related for interbreeding to produce offspring classified within the same species and with the same morphological characteristics.
  2. a category in plant classification, the rank below *genus*, containing closely related, very similar individual plants. Species are ordinarily grouped to form genera. The following are examples of species within a genus (group). Sequoia (Redwood tree) is the genus name and the species (or varieties) that belong to that group are sempervirens, gracilis, pendula etc. These are written along with Sequoia. (E.g. Sequoia sempervirens is one species, Sequoia gracilis is another and so on.)
  3. (L. *species*, particular kind) a group of interbreeding individuals, not interbreeding with another such group, being a taxonomic unit including two names in binomial nomenclature, the generic name and specific epithet, similar and related species being grouped into a genus.

- **specific**
  pertaining to a species.

- **speculum**
  the shiny, coloured region of the *labellum*, as in some Ophrys species.

- **spermatium**
  non-motile male gamete in Rhodophyta.

• **spermatophyta**
(Gr. *sperma*, seed, *phyton*, plant) a major division of the plant kingdom, characterised by reproducing by seed and subdivided into the Gymnospermae and Angiospermae.

• **sphagnum**
any moss of the genus *Sphagnum*, sometimes used as a growing medium for e.g. epiphytes.

• **sphagnum moss**
type of very porous moss found in bogs. Used in the true bog garden, it supplies the acidic conditions necessary for carnivorous plants.

• **sphagnum peat**
the decaying sphagnum moss.

• **spheroid**
spherical with slightly flattened ends.

• **spheroidal**
describing the shape of a pollen grain or spore in which the polar axis and the equatorial diameter are approximately equal.

• **spica**
in botany, a spike.

• **spicate**
(L. *spicatus*, pp. of *spicare*, to furnish with spikes) having the form of a spike.

• **spice**
one of the fragrant vegetable condiments used for the seasoning of food, such as pepper, allspice, nutmeg, ginger, cinnamon and cloves; collectively, such substances as a class, as the trade in spice(s).

• **spiciform**
shaped like a spike.

• **spicosity**
the state or condition of having spikes.

• **spicula**
a small or secondary spike.

• **spiculate**
covered with or divided into small spikes.

• **spicule**
1. a very small, fine spine or in *Opuntioideae* a glochid. A small, sharp point.
2. (L. *spicula*, a small spike) a small, slender, sharp-pointed piece, usually on a surface. A small spike of flowers.

• **spike**
1. multiple cluster of florets directly attached to an unbranched flowering stem.
2. long, narrow, unbranched cluster of *sessile* flowers produced along a central axis.

• **spikelet**

1. a secondary spike, especially applied to grasses, whose flowers are grouped on several small spikes attached to a central axis.

2. a small spike of a large one, a subdivision of a spike as the spikelets of grasses.

• **spina / spinae**

synonym of Spine.

• **spination**

the characteristics and distribution of spines, which may in some cases be distinguishing. features.

• **spindle**

to shoot or grow in a long, slender stalk or body, to form a stem. To grow a long stalk or stem instead of forming flowers or fruit.

• **spine**

1. a sharp outgrowth from the epidermis which may be stout and woody or fine and hair-like. In *Cactaceae,* always arising multiply from an areole and originating as a modified leaf. Spines are clasified into central spines located centrally on the areole and radial spines around the margins of the areole. Spines protect the plant from some predators, albeit inefficiently. In some species, spines act as a drip tip where atmospheric moisture from dews, mists and rain precipitates on the spines and drips off the tip onto the soil near the base of the plant for uptake by the shallow root system. Compare with thorns.

2. a modified leaf, which can be woolly, bristly, hair-like, needle-like, barbed or curved and may be coloured white, red or yellow to black. Nearly all cacti and some succulents have spines. Cacti spines grow from *areoles* and detach easily, but succulent spines grow right from the stem tissue. Although spines protect the soft plants against predators, their main function is to condense moisture so that it drips to the ground.

3. a stem of leaf modified for protection.

4. (ME, *thorn,* spinal column, fr. L. *spinsa,* thorn, spine, spinal column) a stiff sharp-pointed plant process as a modified leaf, leaf part, petiole or stipule.

5. a general word, applied in palynology to long and tapering pointed elements, exceeding 1μm.

• **spinescent**

ending in a spine or sharp point. Tending to by spiny or thorny.

- **spiniform**
having the form of a spine or thorn.

- **spinose**
full of spines. Equipped with spines or thorns.

- **spinule**
small spines, less than 3μm in length.

- **spinulescent**
producing small spines having the form of a small spine.

- **spinulose**
having small spines.

- **spiral**
corkscrew-shaped.

- **spiraperturate**
describing a pollen grain with one or more spiral apertures. Example Eriocaulon aquaticum (Eriocaulaceae).

- **splitter - coloquial**
one who takes a narrow view of a genus, dividing up variant forms into individual species.

- **spongy mesophyll**
the region of ground tissue in a leaf where parenchyma cells are branched and intercellular air spaces are extensive, usually in the lower half of the leaf.

- **sporange**
the case or sac in plants in which the spores are produced or carried.

- **sporangiophore**
1. the stalk which terminates in a sporangium is produced.
2. the umbrella-shaped sporangium-bearing unit of the strobilus, as in *Equisetum.*

- **sporangium**
a spore case.

- **sporangium, sporangia**
structure in which spores are produced.

- **spore**
1. a single, vegetative, reproductive cell formed by lower plants (cryptogams), that does not contain an embryo.
2. a haploid (1N) reproductive cell capable of developing directly into a gametophyte without uniting with another cell.
3. these reproductive bodies are smaller than seeds. A seeds consists of a seed coat surrounding an embryo plant and the cotyledons or seed leaves. A spore consists of a spore coat encasing a minute amount of protoplasm. Spores are produced by almost all flowerless plants such as, Fungi, Algae, Ferns, Lycopodiums and Equisetums. A large Fern can produce 30 million

spores in one season. When the ripe spores of a Fern come into contact with moisture, they form flat, green, heart-shaped bodes called *prothalli* these bear the sexual reproductive organs and, when fertilised, young Ferns are produced.
4. cell which functions in reproduction but not as a gamete.
5. (Gr. *sporos*, seed) any of various small reproductive bodies, often consisting of a single cell, produced by mosses, ferns etc. asexually (asexual spore) or by the union of gametes (sexual spores) They are highly resistant and are capable of giving rise to a new adult individual, either immediately or after an interval of dormancy.
6. a general term for the usually microscopic, unicellular, asexual or sexual reproductive units of cryptogams and fungi.

- **spore case**

the sporangium or covering of the spores.

- **spore shapes**

monolete. Bean-shaped, with a single scar line.

- **sporocarp**

a hard, nut-like structure containing the sporangia in heterosporous ferns.

- **sporoderm**

1. the entire wall of a pollen grain or spore.
2. a general term for spore-like palynomorphs.

- **sporophyll**

1. a modified leaf bearing sporangia.
2. the leaf or leaf-like organ that bears the spores.
3. a spore-bearing leaf.
4. a leaf bearing sporangia.
5. fertile, spore-producing blade as in brown algal order Laminariales.

- **sporophyte**

1. the spore producing, diploid (2N) phase of the life cycle, compare to gametophyte.
2. the stage of the life cycle of ferns and seed plants in which the nonsexual organs or reproduction are borne, as distinguished from *gametophyte*.
3. the usually diploid spore-producing phase of a life history.
4. (Gr. *sporo*, a seed *phyton*, plant) the diploid spore-producing phase in plants with alternation of generations.

- **sporopollenin**

1. an organic polymer that makes up the coat of a pollen grain or spore. Sporopollenin is extremely resistant to degradation.

2. the carotenoid-like polymer comprising the outer wall of pollen grains and certain spores and apparently dinoflagellates and acritarchs. Highly resistant to forms of deterioration other than oxidation.
3. the name given to the acetolysis resistant biopolymers which make up most of the material of the exine.

- **sport**

this means a sudden variation in habit of growth or blossom color from the rest of the plant or others plants of its kind. This is caused by a genetic change that may be accidental or spontaneous or intentionally induced. Therefore, if a pink flower appears on a yellow variety of Mum, it is said to be a sport of that variety. Likewise, bush Roses often give rise to climbing 'sports'. The botanists call such sports, mutants.

- **spreading**

growing outwardly by means of new shoots or advancing outwards by means of rooting from reclining stems. In *Cactaceae*, an array of radial spines projecting obliquely outwards around an areole, but not lying flat against the surface of the plant.

- **sprout**

to shoot forth or begin to grow, as a bud from a seed. To put forth shoot. A shoot of a plant, as from a germinating seed, a rootstock, tuber, runner etc. or from the root (a sucker), stump or trunk of a tree.

- **spur**

1. a usually slender hollow projection formed by the fused bases of the lower petals, often containing nectar, in some flowers or a short fruit-bearing stalk.
2. these are the slender, tubular projections of the petals or sepals of some flowers such as Columbines, Delphinums and Dutchman's Breeches. This word also describes short, woody side shoots growing along the branches of some trees and shrubs such as Apples and Pears. A saccate or tubular extension of the *labellum* or other floral parts in many Orchid species, usually bearing nectar.
3. (AS. *spura, spora*, a spur) a slender, tubelike structure formed by an extension of one or more petals or sepals. Also refers to a very short branch with closely spaced leaves.
4. a usually slender hollow projection formed by the fused bases of the lower petals in

some flowers notably of the genera Pinguicula and Utricularia. A short fruit-bearing stalk.

- **spur-pruning**
  a method of pruning trees, by which one or two eyes of the previous year's wood are left and the rest cut off, so as to leave spurs or short rods.

- **squama**
  a scale or scale-like structure.

- **squamaceous**
  scaly.

- **squamella**
  a small scale.

- **squamiform**
  having the shape, character or appearance of a scale.

- **squamous**
  scaly or scale-like.

- **squamous bulb**
  a bulb in which the outer scales are distinct, fleshy and overlapping. A scaly bulb.

- **squamule**
  a tiny scale.

- **squamuliform**
  having the form or character of a squamule.

- **square tetrad**
  synonym of tetragonal tetrad.

- **squarrose**
  rough with spreading processes, as the whorls of small leaves of various Compositae (daisy family).

- **stalk**
  1. this is the main stem of a plant, which ordinarily supports the leaves, flowers and fruit, as a stalk of wheat or hemp. Or it can describe the *pedicel* of a flower or the *peduncle* of a flower cluster, the *petiole* of a leaf, the *stipe* of an ovary or any other similar supporting organ.
  2. (probably from Dan. *Stilk* Sw. *stjelk* Icel. *stilker*, a stalk) the stem or main axis of a plant, which supports the leaves, flowers and fruit. Any lengthened support on which an organ grows, as the petiole of a leaf or the peduncle of a flower.

- **stalklet**
  a secondary or small stalk.

- **stamen**
  1. pollen-bearing male sexual organ of a flower, consisting of anther and filament.
  2. the male floral organ, bearing the *anther*, which produces pollen.
  3. interior to petals, produces the pollen.

4. (L. *stare*, to stand) the male reproductive organs in flowers. It is situated immediately within the petals and is composed, in most cases, of two parts, the filament and the anther, which is filled with pollen.
5. the male sexual organ supporting and incorporating the anther.

- **staminate**
  1. of a male flower possessing stamens but not a pistil.
  2. equipped with or producing stamens.
  3. male, producing stamens but no *pistils*.
  4. producing or consisting of stamens. Flowers with stamens but not pistils.

- **staminode**
  a sterile stamen.

- **staminodium**
  (L. *stamen*, a thread and Gr. *eidos*, form) a sterile stamen or an organ resembling one.

- **staminody**
  the metamorphosis of various floral organs, such as petals, bracts, pistils etc. into stamens. Compare sepalody, petalody, pistilody.

- **standard**
  a tree or shrub that stands alone, unattached to any wall or support, as distinguished from a cordon or an espalier. A plant that is trained to a single stem in tree form. A shrub, such as a rose, can be grafted onto an erect-growing stem. A tree-like cactus may be produced by grafting a trailing or arching species, such as *Schlumber -gera*, onto a tall or columnar species.

- **standing crop**
  weight or organic material that can be sampled or harvested at any one time from a given area, but may not necessarily include the entire plant. Usually refers to normal harvesting procedures, unless specified, for the particular plant under consideration.

- **starch**
  an insoluble high molecular weight (typically several million kDa) polysaccharide formed from beta-D-glucose units linked together with alpha-1,4' linkages (compare with cellulose). Amylose starches are long unbranched molecules whyle amylopectins are branched polysaccharides with alpha-1,6' linkages at the branch point. Starches can be broken down easily into glucose, as required, by amylase enzymes in the tissues and are the principal food storage substance of plants.

Starch can be recognised by forming a characteristic blue colour with iodine.

- **stele**
  1. the central vascular cylinder in stems and roots where the vascular tissue is located.
  2. the central primary vascular system of the stem and associated tissues. Delimited from the cortex by endodermis and pericycle.
  3. the arrangement of vascular bundles in roots and stems.

- **stellate**
  star-shaped.

- **stellate leaves**
  three or more leaves that surround the stem in a whorl or radiate like the spokes of a wheel or the points of a star.

- **stem**
  1. the main axis of a plant which may carry other parts.
  2. same as axis, a plant axis with leaves or enations.
  3. the main (usually aerial) axis of a tree, shrub or plant, trunk, stalk. The plant part that supports the leaves, flowers or fruits of a plant, as the *peduncle* of a fruit or the *pedicel* of a flower; the *petiole* or leaf-stem.
  4. the portion of a plant axis produced by the shoot apical meristem.
  5. (AS. *stemm*, tree stem) main axis of a plant typically above the soil surface, having leaves or scales and a characteristic arrangement of the vascular tissue.
  6. the main body or stalk of a plant.

- **stem cutting**
  a cutting taken from a portion of stem.

- **stem tuber**
  swollen structures produced by stolons and runners which remain dormant during adverse conditions and later grow into new plants when the conditions become favorable for growth.

- **stem-leaf**
  this is a leaf growing from a stem, a cauline leaf.

- **stemless**
  having no stem, acaulescent.

- **stenopalynous**
  describing plant taxa characterised by only a slight variation in their palynomorphs.

- **stenopetalous**
  having narrow petals.

- **stenophyllous**
  having narrow leaves.

- **stephano-**
synonym of zono-.

- **stephanokont**
with ring of flagella.

- **sterile**
1. incapable of engaging in the process of sexual reproduction.
2. unfruitful, unproductive, not fertile, barren. A *staminate* or male flower, a fern frond without *sori*.
3. incapable of engaging in the process of fertilisation.

- **sterilisation**
the process of making sterile, specifically, the process of freeing from living germs.

- **sterilise**
to disinfect pruning tools to avoid transferring disease. This is normally done by dipping in a 5% solution of household bleach. To kill all living organisms in soil to control pests, done with soil *solarisation* or chemical treatment.

- **sternotribe**
touching the sternum, as certain flowers, especially adapted for cross-fertilisation by external aid, have stamens and styles so arranged as to touch the sternum of a visiting insect.

- **stichidia**
specialised branch bearing tetrasporangia in some Rhodophyta, usually larger than vegetative branches

- **stigma**
1. receptive surface or structure for pollen at apex of style or pistil. One style may bear multiple stigmas, in *Cactaceae* usually 3-20.
2. the part of the *pistil* of a flower, normally on the tip of the *style* or *ovary*, which receives pollen.
3. the portion of the pistil that receives the pollen.
4. (Gr. *stizein* to prick) the upper tip or part of the pistil of a flower receiving the pollen. It is generally situated at the upper extremity of the style.
5. the female organ that receives and is fertilised by pollen.

- **stigmatiferous**
having a stigma.

- **sting**
a sharp, hollow, glandular hair that secretes an irritating or poisonous fluid, as in the nettle.

- **stipe**
1. the stalk that connects *pollinia* and *viscidium* in a *pollinarium* (of Orchids). A stalk or support, as the stalk-like prolongation of the receptacle of a flower, the petiole of a fern leaf

or the stem that supports the pileus of a mushroom.
2. stalk or erect portion as in brown algal order Laminariales
3. (L. *stipes*, stalk) the stalk-like basal part of an ovary or of a fruit such as an achene, the stem bearing pileus in mushrooms and toadstools.
4. the stalk that supports the carpel.

- **stipe bundles**
the vascular bundles of the fern petiole.

- **stipel**
1. a secondary stipule located at the base of the leaflets of a compound leaf.
2. a small secondary stipule at the base of a leaflet.

- **stipiform**
having the form or appearance of a stipe, stalk-like.

- **stipitate**
having or being supported by a stipe.

- **stipular**
having or resembling stipules, as stipular glands.

- **stipulary buds**
accessory buds in the axils of stipules, as in the apple and pear.

- **stipulate**
having stipules, as a stipulate leaf or stalk.

- **stipule**
1. one of a pair of usually leaf-like lateral appendages found at the base of the petiole of many leaves.
2. (L. *stipula*, a stalk, a straw) one of two foliaceous or membranaceous processes developed at base of a leaf petiole, sometimes in tendril or spine form, sometimes fused.
3. a small modified leaf usually found at the base of a leaf.

- **stipuliform**
having the form of a stipule.

- **stirps**
a race or permanent variety of plants.

- **stobiloid**
resembling or relating to a strobile.

- **stock**
the stem, stalk or trunk of a tree or other plant, as distinguished from roots and branches. A stem or root in which a bud or scion is inserted in grafting also, a tree, stem or plant that provides slips or cuttings.

- **stolon**
1. a prostrate basal branch above or below the ground which can root and produce new plants at its tip.

2. a horizontally spreading or arching stem that runs along the ground or just below the surface, which roots at its tip to produce a new plant. Often confused with *runner.*
3. (L. *stole, stolonis*, a twig, shoot) a stem which grows from a stem above the ground, taking root at the tip and ultimately developing a new plant.
4. a horizontal stem that takes root at points along its length to form new plants.

- **stoma**

1. (Pl. *stomata*)- a very small opening or pore through the plant's epidermis, through which repiration and transpiration occurs, surrounded by two guard cells. The stoma is usually capable of active opening and closing through changes in shape of the guard cells. Stomata are found on the plant body of cacti and on any leaves.
2. (NL. fr. Gr. mouth - more at stomach) one of the minute openings in the epidermis of leaves, stems and other plant organs through which gaseous interchange between the atmosphere and the intercellular spaces within these structures occur; the opening together with its associated guard cells and accessory cells.
3. a leaf pore.
4. a pore on a plant's stem or leaf, which through opening and closing, controls the exchange of gases with the outside.
5. a minute prore or opening in the epidermis of leaves, stomata are flanked by two guard cells that regulate opening and closing of the pore and thus regulate gas exchange and transpiration.

- **stomatal closure response**

the response by plants to environmental stress such as drought, which reduces stomatal pore size. This leads to reduced water transpiration rates and restricted access of CO_2 to the leaves.

- **stomatal crypts**

a pit containing several stomata.

- **stomate / stomata**

an opening defined by pairs of guard cells that controls gas exchange and water loss.

- **stomium**

lip cell region of a fern sporangium.

- **stone fruit**

a fruit with a hard endocarp enveloped in a pulp, as the peach, cherry and plume. A drupe or drupelet.

- **stool**

the root or stump of a tree or of a bush, grass, cane etc., which produces shoots each year; also, the cluster of shoots thus produced. The mother plant from which young plants are propagated by the process of layering. *v.* To produce shoots form the root; to form a stool.

- **stopping**

the removal of the growing tip of a plant with the finger and thumb to encourage bushy growth. Also known as 'pinching out'.

- **storage parenchyma**

characterised by large accumulations of storage products such as starch, protein, oil, hemicellulose or water.

- **storage root**

a root modified to store relatively large amounts of food.

- **storied cambium**

fusiform initials aligned with one another.

- **stove plant**

a plant from a warm climate that must be grown in a greenhouse in order to live.

- **strain**

a group of plants within a variety having some unique quality, such as flower colour, drought resistance or high yield.

- **stramineus**

(L. *stramen*, straw) of or like straw, straw-coloured.

- **stratification**

1. a process of exposing sown seeds to low, near freezing temperatures to imitate natural over-wintering as an aid to germination. During the process, chemical inhibitors of germination are broken down.
2. the storage of seeds in warm or cold conditions to break dormancy and aid germination. Some seeds won't sprout right after they ripen, but will die if they dry out. They need to have a moist, dark place for a certain period of time before they will germinate. In nature, this is accomplished as the seeds fall to the ground in the fall or early winter and are covered with falling leaves, which provide darkness and moisture. The winter months provide the low temperatures. Gardeners can mimic this by combining the seeds with sand or sand and peat moss and storing them where they will remain moist and cool. Very small seeds can be placed in between cheesecloth and covered with sand so they may be found easily. Seeds may also be

mixed with slightly moist sand, placed in jars and stored in the refrigerator.
3. process of refrigerating sown seeds to imitate natural wintering as an aid to germination.

- **stratigraphic palynology**
the use of palynomorphs, their identification, distribution and abundance to correlate among sedimentary sequences of any age or to provide chronological control for these sedimentary sequences. Exclusive of environmental palynology, but the study of sedimentary sequences often includes both stratigraphic and environmental palynology.

- **stratum**
a subdivision of a major layer of the sporoderm.

- **streptomycin**
a bactericide used to control several plant diseases, including fire blight.

- **stress**
a term used for the total energy with which water is held in the soil. Expressed in any convenient pressure unit.

- **striated**
(L. *striatus*, grooved) marked by narrow lines or grooves, usually parallel.

- **striato-reticulate**
describing a pattern in which parallel or subparallel muri are cross-linked to form a reticulum in the grooves. The connections between the muri lie on a single level or different levels. Example Gentiana pneumonathe (Gentianaceae).

- **strict consensus**
a method for choosing among several most parsimonious trees generated by a phylogenetic analysis, strict consensus means that only clades that show up in all the most parsimonious trees are recognised.

- **striga**
a sharp-pointed *appressed* bristle or hair-like scale.

- **strigose**
rough with strigae. Covered with sharp and appressed straight and rigid bristles or hairs, as a strigose leaf or stem.

- **strike**
to take root, as of a slip of a plant.

- **striolate**
minutely striate or striped.

- **strobilaceous**
resembling or relating to a strobile. Bearing strobiles.

- **strobile**
  a cone-like structure of *sporophylls*, as in the horsetails and club mosses. A gymnosperm cone or a hop inflorescence.

- **strobilus**
  1. a cone-shaped cluster of spore-bearing leaves.
  2. stem with short internodes and spore-bearing appendages, a cone.
  3. cone, an aggregation of sporophylls on a common axis.

- **stromatolite**
  fossilised, calcareous accumulation of Cyanobacteria.

- **strombuliform (adj)**
  shaped like *Strombus*- a twisted, conical seashell.

- **strophiole**
  of a band or twisted band, a crestlike, corky excrescence about the hilum of certain seeds, derived from the funicle.

- **structure / structurate / structured**
  the internal construction of the pollen or spore wall.

- **struma**
  a cushion-like swelling of or on an organ, especially at one side of the base of the capsule in many mosses.

- **strumose**
  bearing a struma.

- **strychnine**
  a bitter alkaloid poison found in the seeds of the asian tree *Strychnos nux-vomica* native to India, Sri Lanka and Australia. Strychnine has been used as a poison for several hundred years and causes characteristic convulsions.

- **stunt**
  to retard or stop plant growth, usually by accident through exposure to harsh weather or lack of water or nutrients.

- **stupose**
  bearing tufts or matted filaments, like tow.

- **stylate**
  having a persistent style.

- **styliferous**
  style-bearing, bearing one or more styles.

- **s-type tetrad**
  a tetrad in which only one members is fully developed. Example Leucopogon fasciculatus (Epacridaceae).

- **sub**
  a prefix to a descriptor meaning 'almost'.

- **subaxillary**
  situated under an axil, as a subaxillary bud.

- **subclass**
  in plant and animal classification, a taxonomic category of related orders within a class.

- **subcordate**
  in the form of two slightly rounded lobes.

- **suberect**
  synonym of prolate spheroidal.

- **subfamily**
  a taxonomic group of related genera (groups) within a family.

- **subgenus**
  1. a subdivision of a genus into morphologically recognisable groupings of species.
  2. a category in taxonomy of related species within a genus (group).

- **subherbaceous**
  a herbaceous plant that becomes more woody later in the growing season.

- **subirrigation**
  the natural or deliberate irrigation through the control of the water table that will raise water into the root zone.

- **subisopolar**
  describing a pollen grain or spore in which the proximal and distal faces are slightly different. Examples one face is convex and the other is less convex (Banksia, Proteaceae) on one face viscin threads are connected (Oenothera, Onagraceae).

- **sub-layer**
  synonym of stratum.

- **sublittoral**
  below lowest low tide level.

- **submerge**
  (L. *submergere*, to dip or plunge under) to sink or plunge beneath the surface of water.

- **submerged**
  adapted to grow or growing under water. Under water.

- **submersed plants**
  plants growing with their root, stems and leaves completely under the surface of the water.

- **suboblate**
  describing the shape of a pollen grain or spore in which the ratio between the polar axis and the equatorial diameter is 0.75-0.88.

- **suborbicular**
  (L. *sub*, under, below; *orbis*, circle) approximately circular.

- **subpetiolar**
  located under or within the base of the *petiole*, as the leaf buds of the plane tree.

- **subprolate**
  describing the shape of a pollen grain or spore in which the

ratio between the polar axis and the equatorial diameter is 1.14-1.33.

- **subrotundate**
less perfectly orbicular than rotundate.

- **sub-shrub**
a perennial that has a *woody* base, but whose herbaceous shoots die back in the winter.

- **subsidiary cells**
cells adjacent to guard cells that are distinct in appearance from ordinary epidermal cells.

- **subsoil**
the soil which lies beneath that which is cultivated. It is usually hard and infertile. When the ground is being dug up to form a garden, the subsoil shouldn't be brought up to the surface and the upper soil buried. The subsoil can be improved by digging deeply and adding manure, compost and other organic materials.

- **subspecies**
1. a subdivision of a species, recognisable by morphological characterisitics, but not isolated by genetic barriers from other members of the species. Sometimes designated as a *variety*.
2. a taxonomic category applied to specimens which form a distinct group within a species, but are not different enough to be classed as a species in their own right.

- **subspecific**
of the nature of a subspecies, not quite specific.

- **subspheroidal**
describing the shape of a pollen grain or spore in which the ratio between the polar axis and the equatorial diameter is 0.75-1.33.

- **substrate**
the medium on which an organism lives, as soil or rock.

- **substratum**
a term for any layer lying beneath the subsoil or solum.

- **subtend**
to occur below, as a bract below a flower.

- **subtransverse**
synonym of oblate spheroidal.

- **subtropical**
subtropical plants are those which can survive somewhat lower temperatures than genuine tropical plants and even an occasional light frost.

- **subturma**
a group of form-genera of fossil spores or pollen in the turma-system of Potonié.

- **subulate**
  slender or tapering and flattened into a narrow elipse in cross-section, awl-shapedor less cylindrical and tapering to a point.

- **subvarietal**
  varying slightly, having the character of a subvariety.

- **subvariety**
  a subordinate variety. The further and minor modification of a variety, as a strain or line.

- **successiform**
  referring to a phyletic series of pollen types with increasing numbers of apertures, ranging from tricolpate to pantocolpate and pantoporate.

- **succession cropping**
  the method of growing new plants in the space left by the harvested ones.

- **succession of crops**
  the rotation of crops. See **rotation of crops.**

- **succubous**
  overlapping with the anterior edge of one leaf passing beneath the posterior edge of that succeeding it. Opposite of *incubous*; said of leaves.

- **succulent**
  juicy (Latin: *succus* = juice,sap), fleshy plants that store water in specially adapted thickened or swollen organs such as leaves, stems or roots. A plant with thick fleshy leaves and/or stems that are adapted to store water. (These include all cacti, though not all succulents are cacti. see **Cactus.**) Stem succulents, which store large amounts of water in their enlarged stems, are often columnar, round or barrel-shaped. Almost all cacti are stem succulents. Root succulents store water underground in their swollen roots. Even though their stems and leaves may also be fleshy, the main water-storing tissue of these succulents lies in the tuberous roots. Leaf succulents form a large group of varied plants. These plants store water in their thick, fleshy leaves, which wither in drought and swell up again when water is accessible. Many leaf succulents have practically no stems, simply forming *rosettes* of overlapping leaves.

- **sucker**
  these are similar to runners except that the horizontal parts of their stems are below the surface of the soil. Rooted suckers can be dug up and planted elsewhere. Shoots that grow from the understock on which a tree

or shrub is budded are also called suckers. These should be cut off as low down as possible, otherwise they may kill off the named varieties that have been grafted or budded onto the understocks.

- **suckers**
  shoots or small plants arising from the base of a larger plant.

- **suffrutescent**
  somewhat or slightly woody at the base, *subshrubby*.

- **suffruticose**
  having stems that are woody at the base and herbaceous above.

- **suffruticulose**
  (L. *sub*, under, below; *frutex*, a shrub) moderately frutescent; obscurely shrubby; usually woody only basally.

- **sulco**
  of a plant or part with a groove or furrow (Latin *sulcus* = furrow or groove).

- **sulculus**
  an elongated latitudinal ectoaperture not situated at a pole.

- **sulcus / sulci / sulcate**
  an elongated latitudinal ectoap-erture situated at the distal or proximal pole of a pollen grain.

- **sun/shade leaves**
  leaves in the sunnier parts of trees are distinct from more shaded leaves.

- **sunken stomata**
  stomata located below the surface of the epidermis.

- **superior**
  ovary is exposed - above the point of attachment of the perianth and anthers.

- **superphosphate**
  a fertiliser that supplies phosphorus and can force some plants to grow faster.

- **supporting cell**
  cell bearig the carpogonial branch in Rhodophyta.

- **suprafoliaceous**
  inserted on the stem above a leaf, as a peduncle or flower.

- **suprafoliar**
  growing above a leaf.

- **supratectal**
  indicating the position of features, such as spines, on the top of the tectum.

- **surculose**
  producing suckers.

- **surculus**
  a sucker.

- **surface freeze**
  a ground condition in which the moisture found on the surface freezes in cold temperatures, thus causing frost heaves.

- **sustainable use**
  the careful exploitation of natural resources without destroying them.

- **suture**
  1. the seam between two edges, as where *carpels* of a *pericarp* join, marking the line of *dehiscence* or opening.
  2. (L. *suere*, to sew) a seam formed when two parts unite, a seam or line or groove, usually applied to the line along which a fruit dehisces. Any lengthwise groove that forms a junction between two parts.
  3. synonym of commissure.

- **swale**
  (Sw. *sval*, cool) a hollow or depression, especially one in wet, marshy ground.

- **swamp**
  (Sw. *svampig*, swampy) spongy land, low ground filled with water, a wooded area having surface water much of the time.

- **syconium**
  a multiple fleshy fruit composed of a hollow receptacle containing numerous reduced flowers.

- **sylvestris-type**
  bisaccate pollen grains in which the outline of the sacci in polar view is discontinuous with the outline of the corpus so that the grains seem to consist of three distinct, more or less oval parts. Examples Pinus sylvestris, Abies (Pinaceae).

- **symbiosis - (greek: = living together)**
  a balanced partnership in which two dissimilar organisms live together for their mutual benefit e.g. the algae and fungae jointly forming lichens.

- **symmetrical**
  divided into two comparable parts. Having the same number of parts in each whorl, said of a flower.

- **sympetalous**
  having the petals joined at the base. The same as gamopetalous.

- **symplesiomorphy, symplesiomorphic trait**
  s*hared* primitive trait (plesiomorphy).

- **sympodium**
  1. a conspicuous stem that imitates a simple stem, but is made up of the bases of number of

secondary axes that arise successively as branches one from another; as the grapevine.
2. an axial bundle and its leaf and branch trace.

- **syn-**
a prefix indicating the fusion or anastomosis of features.

- **synacmic**
of or pertaining to synacmy.

- **synacmy**
synantheses, simultaneous maturity of the anthers and stigmas of a flower, opposed to heteracmy.

- **synangium, synangia**
a reproductive unit composed of fused sporangia.

- **synantherous**
having the stamens joined together by their anthers, as in the Daisy family, Compositae.

- **synanthesis**
simultaneous anthesis; the synchronous maturity of the anthers and stigmas of a flower; synacmy.

- **synanthous**
a plant having flowers and leaves that emerge at the same time, exhibiting synanthy.

- **synanthy**
the more or less complete union of several flowers that are usually distinct.

- **syncarp**
1. an aggregate fruit, as the blackberry.
2. (Gr. *syn*, together *karpos*, fruit) a multiple or aggregate fruit derived from numerous separate ovaries of a single flower. a collective unit, as a blackberry.

- **syncarpous**
having the character of a syncarp. Having united carpels.

- **syncolp(or)ate**
describing a pollen grain with two or more simple (or compound) colpi the ends of which anastomose at the pole. Example Primula farinosa (Primu-laceae).

- **syngenesious**
having anthers united in a ring, as the stamens of *Compositae*.

- **synrugoidate**
describing a pollen grain with six colpi of which three are long and meeting at one pole and three are short and not meeting at either pole. Example Schisandra grandiflora (Schisandraceae).

- **synsepal**
a floral part formed by the partial or complete fusion of two or more sepals.

- **syntepalous**
flowers in which the tepals are fused.

- **systematics**
  1. field of biology that deals with the grouping and organising of organisms.
  2. the study of the identification (taxonomy) and evolutionary relationships among plant species.

- **tactic**
  of a movement made by a whole organism or by a freely motile part such as a gamete, in response to a directional stimulus which has a direct bearing on the direction of the elicited response.

- **taenia**
  (Gr. *tainia*, a ribbon or tape) formation of ribbon-like structure with little or no differentiation between the leaf blade and stem.

- **taenia / taeniae / taeniate**
  one or more strap-like, more or less parallel strips of exine on the proximal and/or distal sides of the corpus of certain fossil gymnospermous pollen grains. Examples: Striatites, Vittatina.

- **taeninate**
  tapeworm-shaped, long and irregularly contracted.

- **tankage**
  this is an organic fertiliser that leaves an alkaline reaction.

- **tannin**
  a heterogeneous group of astringent phenolic substances, widely distributed in plants and commercially used in tanning, dyeing and preparation of ink.

- **tannins**
  1. phenolic compounds than complex with protein, function in plant defence.
  2. (L. *tannum*, oak bark) complex aromatic compounds some of which are glucosides, possibly giving protection or concerned with pigment formation.

- **tapetum**
  a layer of nutritive cells that lines the pollen sac.

- **taproot**
  1. a primary root growing straight down from a plant and larger or longer than the branching fibrous roots. Often a thick storage organ as in a carrot.
  2. the main, downward-growing root of a plant, which grows deeply and produces lateral roots along its length.

- **tarp**
  this is a waterproof sheet, usually made of polyethylene, used to hold and move bulky yard material such as leaves.

- **taxic homology**
  correspondence in structure between sister groups, for example fronds of ferns and leaves of dicots are taxic homologies, i.e. they are derived structures from the same primitive structure of the common ancestor of these two groups.

- **taxon**
  1. a formally described category of individuals, distinct by some visible characteristics.
  2. (Gr. *taxis*, arrangement) a taxonomic category or unit, as a species or genus.

- **tectum / tecta / tectate**
  the layer of sexine, which forms a roof over the columellae, granules or other infratectal elements.

- **tectum imperforatum / tectate imperforate**
  with a continuous tectum, without perforations. Synonym of eutectum.

- **tectum perforatum / tectate perforate**
  a tectum with perforations smaller than 1μm in diameter. See **punctum.**

- **tectum solidum**
  synonym of eutectum.

- **teeth**
  sharp, pointed triangular structures arising on leaf margins.

- **tegillum / tegilla tegillat**
  synonym of tectum.

- **temperate**
  a climatic zone between boreal and tropical.

- **tendril**
  1. a slender leafless side-shoot, often coiling around other objects, produced by many climbing plants as an aid to climbing and support.
  2. a spirally, coiling, thread-like organ that attaches itself to another body for support, as in the Grapevine or Virginia Creeper. A tendril may be a modified stem, branch, leaf or stipule.
  3. (O.Fr. *tendrillon*, tender sprig) a slender twining or clasping process, modified stem, leaf or part of a leaf, by which some plants climb.
  4. a slender leafless shoot produced by many Nepenthes and other climbing plants as an aid to climbing.

- **tension**
  a term related to soil moisture and the equivalent negative pressure of suction of water in the soil.

- **tenui-**
  a prefix for thin.

- **tenuitas / tenuitates**
  a general term for a thinning, that has been applied to many

different situations in palynology.

- **tepee**
  this is a cone-shaped structure consisting of three or more stakes bound together near the peak used to grow vine crops such as peas, beans and cucumbers.

- **terete**
  1. pencil-like, round in cross-section, cylindrical, but usually tapering at both ends, such as a terete seedpod.
  2. (L. *teres*, round, smooth) nearly cylindrical in cross-section, as stems.
  3. smooth and rounded, circular or nearly so, in section but not necessarily cylindrical.

- **terminal**
  1. borne at the end of a stem, branch or other structure e.g. terminal spine located at the end of a leaf.
  2. growing at the tip of a branch or stem, often applied to a bud, rosette or flower.
  3. borne at the end of a stem.

- **ternatisect**
  cut into three lobes or partial divisions.

- **terracotta**
  a brownish-orange earthenware used to make pots.

- **terrarium**
  an enclosed environment for plants or land animals. A place where such plants or animals are kept alive for study or observation.

- **terrestrial**
  growing in or on the ground. Means a plant which derives its nutrients from the soil rather than atmospheric moisture as do *ephiphytic* plants or the water as do *aquatic* plants. Some Orchids are epiphytes and some are terrestrial. Water Lilies are an aquatic plant.

- **tessellate**
  a surface patterned like a mosaic or pavement. Hence tessellata.

- **tessellation**
  a checkered pattern.

- **testicular**
  in the form of two oblong bodies .

- **testicular**
  in the form of two oblong bodies.

- **tetrad mark**
  the monolete or trilete mark on the proximal face of a spore or, more rarely, a pollen grain.

- **tetrad stage**
  the period during post-meiotic development when the four

microspores or megaspores are united by the presence of a temporary special wall.

- **tetradynamous**
having six stamens, four of which are long, as in the *Cruciferae.*

- **tetragonal tetrad**
a uniplanar tetrad in which all four members are in contact at the centre of the tetrad so that, in the correct orientation, the adjacent walls form a cross. Example Uvariastrum hexaloboi -des (Annonaceae).0000000000-00000.

- **tetragynous**
having four pistils or carpels.

- **tetrahedral tetrad**
a multiplanar tetrad in which each member is in contact with three others, so that the centres of the grains define a tetrahedron. Example Erica (Ericaceae).

- **tetramerous**
1. having the parts in sets or multiples of fours. Often written 4-merous.
2. having four stamens.

- **tetrasporangia**
cell in which diploid (2N) nucleus undergoes meiosis and forms four haploid (N) spores or tetraspores in the Rhodophyta.

- **tetrasporophyte**
diploid phase which produces tetraspores in the Rhodophyta.

- **textural class**
the type or classification of soil material determined by the proportion of sand, silt and clay in its makeup.

- **thalamus**
a receptacle. The end of the flower stalk where the flowers are carried, which is often slightly enlarged.

- **thalliform**
having the form of a thallus.

- **thalloid**
1. resembling or consisting of a thallus.
2. of or resembling a thallus.

- **thallophyte**
1. (Gr. *thallos*, a young shoot *phyton*, a plant) a primary division of plants including all forms consisting of one cell and cell aggregates not clearly differentiated into root, stem and leaf, including bacteria, algae, fungi and lichens.
2. an organism not differentiated into shoot and root, the cells of which have a cell wall. originally used for bacteria, fungi and algae.

- **thallus**
1. the body of mosses, lichens,

algae and fungi, plants without leaves, roots or stems.
2. term given to the body of a lower plant which has no recognisable shoot, root, leaf regions organism's body, not differentiated into roots, stems, leaves.
3. (Gr. *thallos*, a young shoot) a plant body that lacks differentiation into distinct forms of stems, leaves, roots and does not grow from an apical point.

- **theca**
a case or sac, it's generally the same as a capsule.

- **thermometer**
an instrument used to measure temperature.

- **thermotropic**
plant movement that is determined by temperature.

- **thermotropism**
bending, growth or movement of a part under the influence of a heat source.

- **therophytes**
(Gr. *theros*, summer;, *phyton*, a plant) an annual plant that overwinters as a seed.

- **thigmotropism - or- haptotropism**
a growth response to a touch stimulus.

- **thorn**
1. a sharp woody outgrowth from the epidermis which unlike the spines of *Cactaceae* contains vascular tissue and arises from the underlying tissues (phylogenetically as a modified stem). Because of this, breaking a thorn off the plant damages the surrounding tissue.
2. a sharp, stiff, woody, spine, which is ordinarily on a plant's branch or stem, such as on a Rose.
3. (ME., fr. OE. akin to OHG. *dorn*, thorn, ON. *thorn*, Goth. *thaurnus*, thorn, Skt. *trna*, grass, blade of grass) a sharp rigid process on a plant; specif., a short, indurated, sharp-pointed and leafless branch developed from a bud in a manner typical to a leafy branch.

- **throat**
1. the visible portion of a flower between the limb and tube.
2. the mouth or orifice of a corolla or calyx, being the place at which the tube joins or expands into the limb.
3. (Gr. *drossel*, the gullet) term applied to an expanded part of a corolla tube just below the lobes.

- **thylakoid**
photosynthetic lamella or sac.

- **thyrse**

a flower cluster with a central spike that has side branches also clustered with flowers, such as a Horse chestnut and lilac.

- **thyrseus**

(Gr. *thyrsos*, wand) a panicle-like inflorescence consisting of a slender indeterminate main axis with lateral axes determinate, i.e., cymose.

- **thyrsoid**

resembling a thyrse.

- **tigella**

a short stem.

- **tigellate**

having a short stalk, such as that of a bean.

- **tillage**

the operation or art of tilling land or preparing it for seed and keeping the ground free from weeds that might impede the growth of crops, cultivation, culture, husbandry. Tillage includes manuring, plowing, harrowing and rolling land or whatever is done to bring it to a proper state to receive the seed and the operations of plowing, harrowing and hoeing the ground to destroy, weeds and loosen the soil after it is planted.

- **tiller**

(OE. *telgor, telgra*, branch, twig, shoot) sprout, stalk, especially one from the base of a plant or from the axils of its lower leaves.

- **tilth**

the condition of a soil in respect to its ability to support the growth of a specific type of plant.

- **tissue culture**

a method of asexual propagation used by commercial growers to produce clones of a particular plant in large quantities. It produces marketable plants faster than growing them from seed. Meristem cells are grown in nutrient solutions in laboratory flasks until they have recognizable roots and leaves. They are then transplanted into a suitable potting medium. The process is widely used in commercial orchid growing and is sometimes called meristem culture.

- **tissue culture**

the use of specialised methods to mass produce plants starting with small amounts of plant tissue.

- **tissues/tissue systems**

groups of cells that share a similar function, such as transport (vascular tissue) or protection (dermal tissue).

- **tomentose**
  1. covered with thickly matted, woolly hairs.
  2. (L. *tomentum*, down) densely covered with short, matted hair.
  3. covered in tomentum.

- **tomentum**
  1. covered with matted woolly threads. Adj. *tomentose*- a surface covered with tomentum. a covering composed of short, thickly matted, woolly hairs.
  2. matted wolly threads on the surface of leaves and stems.

- **toothed margin (leaves)**
  margin having projections or serrations with pointed apices, indented less than one quarter of the distance to the midvein or long axis of the leaf, compare entire margin and lobed margin leaves.

- **top soil**
  a term for a presumed fertile soil. The surface plow layer of soil. The upper 3 feet of soil. The A-horizon or surface soil. The surface or upper part of the soil.

- **top-dressing**
  a material, such as compost or manure, that is applied to the surface around the plant to aid in drainage, decrease erosion, prevent moisture loss and keep weeds down, as well as dress up the plant.

- **topiary**
  clipped or cut into ornamental shapes. Topiary work is the clipping and trimming of shrubs and trees into regular or whimsical shapes.

- **topography**
  the shape of the ground's surface like hills, valleys, plains and slopes.

- **torose**
  knobbed. Cylindrical with regular bulges.

- **tortuous**
  twisted.

- **torulous (adj)**
  knotted, like moniliform, but with uneven bulges.

- **torus / tori**
  an arcuate invagination or protuberance of the exine more or less paralleling the laesura of a spore in the interradial area.

- **trace elements**
  very small amounts of a chemical element found in plant tissue.

- **tracheary**
  of conducting tissues containing tracheids and vessels.

- **tracheary element**
  a conducting cell of the xylem, characterised by an elongated

shape and lignified secondary cell wall.

- **tracheid**
  1. a basic primitive unit of the xylem (fluid-conducting tissue), consisting of a single elongated cell with pointed ends and a thick, pitted secondary cell wall composed of cellulose thickened with lignin. The mature tracheid is dead and empty of cytoplasm. Tracheids are found in all vascular plants and are the only conducting elements in gymnosperms and ferns.
  2. a water conducting and supportive cell type of xylem composed of long, thin cells with tapered ends and walls hardened with lignin.
  3. a tracheary element that lacks perforations plates, water flows from between tracheids through pits.

- **tracheophyta**
  1. (LL. *trachia*, windpipe and Gr. *phyton*, plant) a division of plants comprising green plants with a vascular system that contains tracheids or tracheary elements, being the Pteridophyta and Spermatophyta, commonly called vascular plants.
  2. plants with true vascular tissue, i.e. xylem and phloem; compare to leptoids and hydroids.

- **trachycarpous**
  having rough fruit.

- **trachyspermous**
  rough-seeded.

- **training**
  the art or operation of forming young trees (or any other plant) to a wall or espalier or causing them to grow in a desired shape.

- **transformational homology**
  an evolutionary series of character states with no significant breaks (from the plesiomorphic ancestor to the apomorphy in the descendant).

- **transgenic plant**
  a genetically modified plant with a new gene (transgene) that may impart a new function. For example, transgenic *Bt* corn can produce its own insecticide.

- **transition zone**
  intercalary meristem in region between stipe and blade where active cell division occurs in the brown algal order Laminariales.

- **trans-pecos**
  of territory in Texas west of the Pecos River.

- **transpiration**
  loss of water vapour through the plant's stomata. C4-metabolic adaptations in the

Cactaceae and other succulents allow transpiration to be minimised.

- **transpiration**
the loss of watery vapour from the surface of foliage.

- **transpire**
to exhale watery vapour.

- **transversal furrow**
synonym of endocolpus.

- **transverse**
across or at right angles to the vertical.

- **transverse**
synonym of oblate.

- **transverse section**
cross section, a section perpendicular to the longitudinal axis of the plant organ.

- **trap**
1. a structure designed to catch and detain animal prey.
2. a leaf modified to trap insects to supply the plant with nitrogen.

- **trapeziform**
in the shape of a four-sided figure with none of the edges being symmetrical.

- **traumotropism**
bending, growth or movement of a part as a response to wounding.

- **tree**
1. a plant with a single distinct trunk between the ground and first branch.
2. a perennial plant that grows from the ground with a single, normally tall, woody, self-supporting trunk or stem and an elevated crown of branches and foliage (or only foliage as in the palms). An herb or shrub growing naturally in or trained into the form or a tree.

- **tree ring**
a circle of cells just under the bark. The width of the ring indicates the amount of lateral growth the tree has undergone in any given year.

- **tree-leaved**
having three leaves or leaflets, as many species of *Trifolium* also called trifoliate or trifoliolate.

- **tri-**
a prefix for three.

- **triandrous**
having 3 stamens.

- **triarch**
consisting of three, for example a triarch stele has three lobes.

- **tricarpellary**
having three, usually fused, carpels.

- **trichoblast**
simple or branched filament at apex of the red family Rhodomelaceae.

- **trichogyne**
receptive elongation of the female reproductive structure in the Rhodophyta where male gametes become attached.

- **trichome**
a hair-like structure growing at one end from the plant's epidermis.

- **trichomes**
cells that project from the surface of the epidermis.

- **trichosclereids**
hair-like, similar to a fibres, except branched.

- **trichothallic growth**
growth from cell division at the base of a filament or group of filaments as in the brown algal order Desmarestiales.

- **trichotomocolpate**
describing a pollen grain with a three-armed colpus. Example Trapa natans (Trapaceae).

- **trichotomosulcate**
describing a pollen grain with a three-armed sulcus. Example Elaeis guineensis (Palmae).

- **tricolpate, tricolporate, triporate**
describing pollen grains with three ectocolpi, three compound apertures or three pores. See **monoaperturate, pantoaperturate, zonoaperturate.**

- **tricostate**
three-ribbed.

- **tridactylate**
literally, three-fingered, with three narrow lobes.

- **tridentate**
forked into three points.

- **trifoliate**
a compound leaf with three leaflets. Having three leaves.

- **trifurcate**
divided into three branches of forks.

- **trigamous**
having three sorts of flowers in the same head, male, female and hermaphrodite.

- **trigonous**
1. with a three-cornered, triangular cross section.
2. trigonal, having three prominent longitudinal angles, as a style or ovary.
3. of triangular cross-section.

- **trijugate**
having three pairs of leaflets.

• **trilete mark**
the triradiate mark of a trilete spore.

• **trilobate**
three-lobed, such as a leaf.

• **trimerous**
having the parts in three's; often written 3-merous.

• **trimorphism**
the development of three distinct forms of flowers, leaves or other parts upon the same plant or upon plants of the same species.

• **trimorphous**
pertaining to or characterised by trimorphism; having three distinct forms.

• **trioecious**
having male, female and hermaphrodite flowers, each on different plants.

• **tripetaloid**
appearing as if there are three petals, as a tripetaloid perianth.

• **tripetalous**
three-petalled.

• **tripinnate**
threefold pinnate, as when leaflets of a bipinnate (divided twice) leaf are themselves pinnate.

• **tripinnatisect**
parted nearly to the base or midrib in a tripinnate manner, such as a leaf.

• **triploid**
1. having three complete sets of chromosomes.
2. (Gr. *triplus*, triple) having or being a chromosome number three times the monoploid number.

• **triprojectate**
describing projectate pollen with three projecting arms on which the apertures are situated, as for example in the Triprojectacites group of fossil pollen grains. Example Aquila-pollenites.

• **triqueter**
three-edged.

• **triquetrous**
three-angled, three-winged. Having a triangular cross section.

• **tristachyous**
three-spiked.

• **tristichous**
arranged in 3 vertical rows or ranks.

• **tristylous**
three-styled.

• **triternate**
a leaf divided into three parts, each of which is divided into

three parts and then again divided into three leaflets, each complete leaf having 27 leaflets.

- **trochlearate**
pulley-shaped, in the form of a pinched circle resembling a figure of eight.

- **tropical**
a climatic zone surrounding the equator.

- **true indusium**
an epidermal outgrowth protecting the sorus.

- **trunk**
(OFr. *Trone*,fr. L. *truncus*, a trunk, stock, stem) the main stem or body of a tree, considered apart from its roots and branches.

- **truss**
a compact terminal flower cluster or fruit cluster, such as the Lilac or Tomato.

- **tryma**
a nut-like *drupe* with the outer part of the seed covering fleshy, leathery or fibrous and with a stony endocarp (inner shell), as in the walnut and hickory nut.

- **tubate**
trumpet-shaped.

- **tube**
1. a narrow flower structure formed by fusion of the tepals.
2. any hollow, elongated body or part of an organ. This is especially applied to a *gamopetalous* corolla or *gamosepalous* calyx and also to a united circle of stamens.

- **tube nucleus**
located within the tube cell.

- **tuber**
1. a thick, fleshy underground modified root or stem section used for storage of food or water, as in a potato.
2. a tuber is a thickened, usually underground, food-storing organ (like a corm) that lacks both a basal plate and tunic-like covering, which corms and bulbs have. Roots and shoots grow from growth buds, called 'eyes', on the surface of the tuber. Some tubers, such as caladiums, diminish in size as the plants grow and form new tubers at the eyes. Others, such as tuberous begonias, increase in size as they store nutrients during the growing season and develop new growth buds at the same time. Tubers may be shriveled and hard or slightly fleshy. They may be round, flat, odd-shaped or rough.
3. a swollen underground stem, such as a potato.
4. (L. *tuber*, a swelling or knob on plants) the short, thickened

fleshy food-storing portion of an underground stem with many surface buds, generally shaped like a rounded protuberance.
5. a modified stem used to store food or water by a plant; the hibernating cell>-mass of certain Australian sundews etc.

- **tubercle**
small, rounded prominence or knob-like protusion from the surface. In *Cactaceae* a roughly pyrimidal protrusion from the stem surface, with an areole at or near its summit. Hence adj. *tuberculate.*

- **tuberculate**
having tubercles. Covered with knobbly projections.

- **tubercules**
silica deposits on the stem ridges, as in *Equisetum.*

- **tuberoid**
a thickened, underground storage organ with stem and root tissues.

- **tuberous**
resembling a tuber, bearing tubers.

- **tuberous root**
1. a generally undistended fibrous root system with thickened fleshy sections scattered along its length. Not a taproot or caudex.
2. tuberous roots look like tubers, but are actually swollen, nutrient-storing root tissue. During the growing season, they put out fibrous roots to take up moisture and nutrients. New growth buds or eyes, form at the base of the stem where it meets the tuberous root. This area is called the crown. To increase tuberous roots, cut off a piece of the root with a section of the crown containing an eye. Day-lilies, dahlias and foxtail lilies have tuberous roots.

- **tubiform**
trumpet-shaped.

- **tubular**
in the form of a hollow cylinder. Forming a hollow nearly true cylinder.

- **tubuliflorous**
when all the flowers of a head have tubular corollas, such as in the Campanulaceae and the Compositae.

- **tubulose**
consisting of or resemblinng, small tubes.

- **tubulus**
a general term for a small channel, applied in palynology to a channel through the nexine.

- **tula / tulae / tulate**
  a sexinal inflation in gymnospermous pollen at the end of the axis of the distal sulcus or leptoma. Example Ovalipollis.

- **tump**
  to form a mass of earth or a round embankment.

- **tunic**
  a loose, outer covering or skin surrounding some corms and bulbs. Examples are the Onion, Tulip and Crocus.

- **tunica**
  the outer layer(s) that divide only anticlinally.

- **turgid**
  1. swollen by some internal force, usually by fluid uptake.
  2. swollen, bloated beyond its natural or regular state by some internal agent or expansive force often said of an enlarged part of the body.

- **turgor pressure**
  outward force exerted on a cell wall by water taken up into the cell by osmosis.

- **turion**
  1. (L. *turio*, shoot) young scaly shoot budded off from underground stems, detachable winter bud used for perennation in many aquatic plants.
  2. a hibernating bud produced by certain aquatic plants from which growth emerges in the following Spring.

- **turma / turmae**
  an artificial suprageneric grouping of form-genera of fossil spores and pollen.

- **turriform**
  biform sculptural elements consisting of a capitate basal portion surmounted by a sharply contracted distal spine. Example Dibolisporites.

- **tussock**
  a tuft or small hillock of growing grass.

- **twig**
  (AS. *twigge*, twig) a small shoot or branch of a tree or other plant.

- **type**
  the permanently preserved specimen or specimens on which the species or sub-species description is based.

- **type locality**
  the place where the type specimen was collected.

- **ulculus / ulculi / ulculate**
  a rounded ectoaperture not situated at a pole. Example Poaceae. See **ulcus**.

- **umbellate**
  1. arranged in umbels, bearing umbels, such as umbellate plants, flowers or clusters.
  2. bearing umbels, consisting of umbels, forming an umbel or umbels.

- **umbelliferous**
  bearing umbels.

- **umbelliform**
  resembling an umbel.

- **umbellulate**
  provided with or arranged in umbellules or umbellets.

- **umbellule**
  a secondary umbel of a compound umbel.

- **umbilicus**
  a scar or depression on the summit of a fruit after the floral parts are shed.

- **umbonate**
  bossed, round in shape, with a rounded projection in the centre.

- **umbraculiform**
  umbrella-shaped.

- **uncinate**
  terminated by a sharply curved hook.

- **uncinate- uncate**
  terminated by a sharply curved hook. Hence uncinatus-with hooked spines.

- **under glass**
  grown in the greenhouse or, in the case of a cutting, with a jar or cloche placed over it.

- **underbrush**
  small trees and/or shrubs growing under large trees in a wood or forest; also called brush or undergrowth.

- **undershrub**
  a shrubby plant that barely grows as large as a shrub, a very small shrub.

- **unequal**
  asymmetrical.

- **ungui**
  claw-like. Hence unguispinus-with claw-like spines.

- **unhumidified**
  said of organic matter before its decomposition into humus.

- **uniaxial**
  having axis composed of one single filament.

- **unicellular trichome**
  consists of one cell.

- **uniflorous**
  having only one flower.

- **unifoliate**
  one-leaved, having one leaf.

- **unifoliolate**
having a leaf that is compound in structure, but has only one leaflet, such as the orange tree.

- **unilocular**
produced in a single cavity or cell as in the unilocular sporangium of the brown algal order Ectocarpales. Single-chambered.

- **uniparous**
having one branch.

- **uniplanar tetrad**
a tetrad in which the individual members lie more or less in one plane.

- **uniseriate rays**
rays that are one cell wide.

- **unisexual**
1. said of a flower containing the organs of one sex,stamens or pistils, but not both, diclinous.
2. having either male or female sex organs but not both.
3. (L. *unus*, one *sexus*, sex) of one or other sex, staminate or pistillate only, but not both.

- **upper pitcher**
the form of Nepenthes pitcher found on the upper portion of the plant.

- **upright cells**
elongated perpendicular to the ray.

- **urceolus**
any urn-shaped organ or part of a plant.

- **utricular**
pertaining to, containing or resembling a utricle.

- **utricularia (g)**
an extremely widespread genus of over 200 species including aquatic, epiphytic and terrestrial species, bladderwort.

- **utriculiferous**
bearing or producing utricles.

- **utriculiform**
having the form of a utricle.

- **utriculoid**
having the form of a bladder.

- **uva**
a succulent fruit having a central placenta, such as a grape.

- **vacuole**
membrane bound organelle that typically occupies a large volume of the cell cytoplasm.

- **vagans**
wandering.

• **valida**
strong, healthy, well developed.

• **vallecular canal**
1. air-filled canals in the cortical (cortex) tissue of some sphenophytes that alternate with the vascular bundles.
2. a canal beneath a stem groove.

• **vallum / valla**
synonym of murus.

• **valva / valvae / valvate**
radial thickenings in the areas beyond the ends of the laesurae of trilete spores. Example Triquitrites tribullatus.

• **valve**
1. a segment into which a dry fruit dehisces.
2. one of the segments into which a capsule or legume splits at maturity or which opens like a lid, as in the barberry.
3. opposite faces of diatom frustule or cell wall.
4. a segment into which a dry fruit dehisces.

• **variegated**
with patches or stripes of different colours, often yellow or reddish. Hence variegata- of a plant exhibiting variegation. Varied in colour, unevenly marked with contrasting colours.

• **variegation**
the combination of two or more colours in the leaves, petals or other parts of plants. Being variegated.

• **vascular**
of tissues containing vessels for conveying fluids and minerals. Referring to a plant's circulatory system, which consists of or relates to, the specialised conducting tissues of plants-xylem and phloem, which circulates sap.

• **vascular cambium**
1. a lateral meristem that produces secondary vascular tissue in stems and roots.
2. a sheet-like meristem that produces secondary xylem and secondary phloem.
3. lateral meristem that forms the secondary tissue and is located between the xylem and phloem.

• **vascular plants**
these are plants in which the structure is made up in part of vascular tissue or vessels. They include the flowering plants (angiosperms), cone-bearing plants (gymnosperms, such as conifers, ginkgos and cycads) and ferns and fern allies (spore-producing). Nonvascular plants include fungi, algae, lichens, mosses and liverworts.

- **vascular tissue**
tissue composed of conducting cells, i.e. xylem and phloem.

- **vascular tissue system**
tissues derived from the procambium or vascular cambium that transport water and photosynthate.

- **vasculate**
vase-shaped, olliform.

- **vegetable**
normally, an herbaceous, cultivated plant used for food, such as potatoes, spinach, peas etc.

- **vegetative**
1. growth of leaves and stems as opposed to flowers. A method of non-sexual reproduction by means of offsets, cuttings.
2. having the power to produce or support growth in plants, such as the vegetative properties of soil.
3. concerning the growth of leaves as opposed to flowers; concerning non-sexual reproduction by means of cuttings, grafting etc.

- **vegetative frond**
frond lacking sporangia.

- **vegetative growth**
non-flowering, usually leafy growth.

- **vegetative propagation**
propagation by asexual means, such as stem and leaf cuttings, layering, root division or bulblets. Leaves of African Violets can be used as cuttings. Roses and fruit trees can be propagated by stem cuttings. Irises can be increased by dividing their roots, Lilies can be increased by bulblets and Strawberries can be layered.

- **vein**
1. a strand of vascular tissue in a flat organ such as a leaf.
2. a fibrovascular bundle at or near the surface of a leaf, a rib.
3. a vascular bundle in a leaf.
4. a rib of tissue, usually in a leaf.

- **velum**
the membranous flap covering the sporangium, as in *Isoetes.*

- **velum / vela / velate**
a feature of a monosaccate pollen grain in which the saccus is convoluted. Example Tsuga (Pinaceae).

- **velutinous**
velvety, covered with thick, silky hairs.

- **venation**

1. the arrangement of veins within a structure.
2. the arrangement of veins in the blade of a leaf.
3. the arrangement of veins within a leaf.

- **venous**

having many veins, such as a venous leaf.

- **ventral**

belonging to the front or inner surface, the opposite of dorsal.

- **ventricose**

1. distended or inflated.
2. swollen or distended on one side.
3. distended, inflated.

- **vermiculate**

worm-shaped.

- **vermiculate**

worm-shaped. A general descriptive term used to describe winding features.

- **vernal**

(L. *vernalis*, of the spring) belonging to the spring, appearing or occurring in spring, of the spring season.

- **vernation**

1. the arrangement of leaves within a bud.
2. the arrangement of leaves within a leaf bud.

- **verruca / verrucae / verrucate**

a wart-like sexine element, more than 1μm wide, that is broader than it is high and is not constricted at the base. Example Plantago (Plantaginaceae).

- **verrucose**

1. of a surface covered with warty protrusions.
2. with a warty or bumpy surface.a (L. *verrucosus*, full of warts, warty) warty, having little warts or wartlike growth on the surface.

- **verrucose adj.**

synonym of verrucate. Resembling a wart.

- **verruculose**

minutely verrucose.

- **verticil**

1. a whorl, as leaves or flowers that are disposed in a circle or ring around an axis.
2. (L. *verticillus*, whorl) an arrangement of leaves, flowers, inflorescences or other structures which surround the stem in a circle upon the same plane about the same point on the axis.

- **verticillaster**
a form of *inflorescence* in which the flowers are arranged in an apparent whorl, consisting of a pair of opposite *axillary*, usually *sessile*, *cymes* or clusters, as in many of the Mint family, Labiatae.

- **verticillastrate**
bearing or arranged in verticillasters.

- **vesiculate**
synonym of saccate.

- **vespertine**
opening in the evening, such as a flower.

- **vessel**
a long tube of vessel elements connected by perforation plates.

- **vestibulum / vestibula**
a separation between layers of the exine forming a cavity between the inner and outer pores. For example Betula (Betulaceae).

- **vestiture**
(L. *vestitus*, grament) that which covers a surface, as hairs, scales etc.

- **vicariance**
separation of a continuously distributed ancestral population or species into separate populations, due to the development of a topographic or ecological barrier.

- **villous**
1. with long soft hairs.
2. having long, soft, individual hairs.
3. (L. *villosus*, hairy) pubescent, shaggy, covered with fine long hairs, but the hairs not matted.

- **villus**
one of the long, straight and soft hairs that sometimes cover the fruit, flowers and other parts of plants.

- **vine**
these are climbing plants with woody or herbaceous stems that climb, twist, adhere or scramble over other taller objects (shrubs, trees, walls, fences, arbors, porches, gazebos etc.) in order to reach sunlight. Vines can climb by tendrils, aerial roots, twining stems, twining leaf-stalks, adhesive disks and hooks. Vines that climb under their own ability are called self-clinging. Some plants used as vines merely sprawl and must

be trained or secured with ties to an upright support, such as a trellis.

- **violacea**
  violet coloured.
- **viperina**
  snake-like.
- **virgin soil**
  soil that hasn't been significantly disturbed from its natural environment.
- **viridescens**
  becoming green.
- **viridiflora**
  with green flowers.

- **virus**
  a simple parasitic micro-organism consisting of little more than a package of genetic material (DNA or RNA) in a protein or lipid envelope, capable of invading the host's cells and replicating only by harnessing their biochemical machinery. Some viruses carry several genes for synthesis of virus-specific proteins. Others (e.g. potato spindle tuber virus) consist of a circle of DNA which is too small to carry any structural genes.

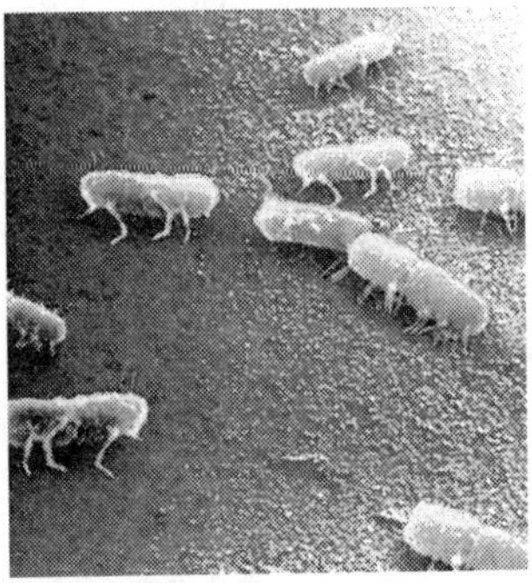

- **viscidium**
  the sticky portion of the *rostellum* which is connected to pollinia.
- **viscin thread (pl. viscin threads)**
  a general botanical term applied in palynology for an acetolysis resistant, sporopollenin thread arising from the exine of a pollen grain, usually from the proximal surface. Example Oenothera (Onagraceae).
- **vista**
  a view of prospect, especially through an avenue, as between rows of trees. The trees or other things that form the avenue.

- **viviparous**
  1. germination that takes place while the new plant is still attached to the parent plant.
  2. (L. *vivus*, alive, *parere*, to produce) producing young alive rather than in eggs, as in most mammals, multiplying by vegetative means such as buds or bulbils in the position of flowers.

- **vulpinus**
  fox-like.

- **wall pressure**
  inward force exerted by a cell wall on the cell contents by virtue of the elastic properties of the cell wall and by pressure from surrounding cells.

- **water farming**
  the cultivation of plants in water, instead of soil. Hydroponic gardening.

- **water logged**
  soil that is completely saturated with water. Water-logged soil has little or no oxygen. The roots of many plants will rot in water-logged soil.

- **water ratio**
  the portion of the total bulk volume of soil that's filled with water.

- **water requirement**
  a term used for units of water required by plants during growing season. Knowledge of how much water a plant needs helps prevent both over and underwatering.

- **water soluble fertiliser**
  an organic or chemical fertiliser which can be dissolved in water, such as manure. Water soluble fertiliser may be applied through a watering system or sprinkled in with a watering can.

- **water table**
  the upper limit of the part of the soil or underlying rock material that is completely saturated with water. A high water table might limit the plants that can be grown successfully.

- **weed**
  a plant growing in the wrong place.

- **wet soil plants**
  plants, such as Willow, Cranberry and Japanese Iris, which can endure a large amount of moisture in the soil.

- **wet weight**
  same as fresh weight; weight of plants after the outer surface covering of water has been removed. Wet weight is not a reliable measurement since methods to prepare plants prior to weighing vary considerably.

- **wet woodland**
  a wooded area having surface water some of the time, for intermittent short periods.

- **wetting agent**
  this is material that reduces the amount of water required by increasing water penetration into the soil. This also can be a material added to pesticide sprays so that they spread easily over the plant surface.

- **whorl**
  1. where three or more organs, such as leaves or petals, arise from the same point and form a circle around it.
  2. (AS. *hweorfan*, to turn) circle of flowers, parts of a flower or leaves arising from one point; verticil.
  3. a ring of leaves etc. around a stem.

- **whorled**
  1. of leaves or other parts, not alternate or opposite but placed in a circle around the stem.
  2. when three or more leaves are arranged at the same level on a stem, typical of such plants as hydrilla.

- **wide row planting**
  the method of sowing seeds, usually of the same plant, in multiple rows. For example, two or three lines of beans can be planted together, often on a raised row, with deep, wide parallel channels for irrigation.

- **wind break**
  something to break the force of the wind, such as a hedge, fence or a row of evergreen trees. Any shelter from the wind.

- **wind burn**
  a withering or blasting of foliage caused by dehydration due to over exposure to wind. Young foliage is especially susceptible. Foliage seriously affected by wind burn won't recover and even though new leaves may

appear, the plant is likely to be weakened.

• **window**

1. a translucent portion of the leaves of some succulent plants which allows light to penetrate to the photosynthetic layers.
2. this is a layer of semi-transparent cells, containing a calcium oxalate, at the tip of a *succulent* leaf. The window protects the *chlorophyll* in the plant body from being destroyed by harsh sunlight by diffusing the light before it reaches the chlorophyll, thus enabling *photosynthesis* to take place.

• **wind-rock**

the destabilising of a plant's roots by the wind.

• **wing**

a dry membranous expansion or thin extension of any kind, such as that on certain seed capsules or stems, such as the winged seeds of Maple Trees (*Acer*). Also, one of the two side petals of certain flowers.

• **wood ashes**

this is a fertiliser containing potash.

• **woody**

1. plants with branches of hard, woody fibres, that persist, unlike soft-stemmed *herbaceous* plants. A semi-woody stem contains some softer tissue and may be only partially persistent. A shrub is an example of a woody plant; its leaves may die off in the winter, but its branches do not.
2. (ME. *woddy, wody*, fr. *wode*, wood + -y) of or containing wood or wood fibres, consisting mainly of hard lignified tissues.

• **woolly**

thickly covered with long hairs.

• **wrack**

tangled mass of seaweeds on the beach.

• **wrinkled**

creased or folded up irregularly in every direction.

• **xeranthemoides**

with a dry appearance.

• **xeriscape**

an arid landscape or water-conserving garden in which xerophytic plants grow. A desert garden. A landscape designed with drought-tolerant plants.

• **xeromorphic**

structural features of plants, xerophytes, adapted to dry conditions. Characters that are advantageous is dry environments.

• **xerophilous**
preferring dryness, describing plants that are characteristic of, tolerant of or thriving in the presence of dry climates.

• **xerophyte**
1. a plant adapted to dry conditions.
2. this is a plant that is adapted to living and growing with limited water because their formation provides water storage or limits *transpiration*, such as cactus and succulents.
3. a plant with xeromorphic features.

• **xylem**
1. the main upward water-conducting tissue in vascular plants characterised by tracheary elements. Xylem may also be a supporting tissue, especially secondary xylem (wood).

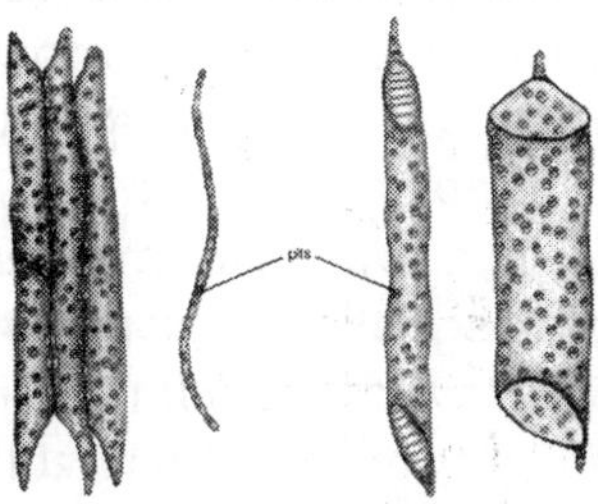

2. tissue through wich most of the water and dissolved minerals utilised be a plant are transported.
3. the water-conducting tissue of plants.
4. (Gr. *xylon*, wood) woody tissue that is part of the water-transporting system in plants, consisting of lignified tracheids or vessels and which also acts as a supporting tissue.

• **xylotomist**
an authority on wood and its composition.

• **yeast**
a unicellular Ascomycete that typically multiplies by a budding process, although some yeasts (e.g. Sacharomyces pombe) multiply by symmetrical division. Sacharomyces species are important mediators of the commercial fermentation of carbohydrates to alcohol.

• **zephyr**
of the west.

• **zinc**
this is a trace element that is usually present in most soils.

• **z-layer**
synonym of exintine.

• **zona- / zonate**
prefix indicating ring-like.

• **zona / zonae / zonate**
a thin outer structure of a spore that projects at the equator, but does not extend over the distal

face or proximal face. Example Cirratriradites saturni.

• **zonal**
relating to the form of a zone. Relating to a major soil group marked by the well-developed characteristics that are determined by the action of climate and/or organism.

• **zonasulculus**
a ring-like sulculus around the equator. Example Nymphaea violacea (Nymphaeaceae). See **meridionosulcus**.

• **zonasulcus**
a meridional ring-like sulcus perpendicular to the equator. Example Laurelia novaezelandiae (Monimiaceae).

• **zonatus**
with concentric markings.

• **zone**
1. a distinct geographic region exhibiting characteristics distinct from neighbouring regions.
2. this is an area that differs considerably from adjacent areas.
3. a distinct region esp. of the interior of a pitfall trap, exhibiting characteristics distinct from neighbouring zones.

• **zonorate**
1. describing a pollen grain with a continuous endoaperture (os) around the equator.
2. Synonym of endocingulum.

• **zoochlorella**
symbiotic green-pigmented algal cell in tissue of animal host.

• **zoospore**
flagellated, asexual reproductive cell.

• **zooxanthella**
symbiotic golden-pigmented algal cell in tissue of animal host.

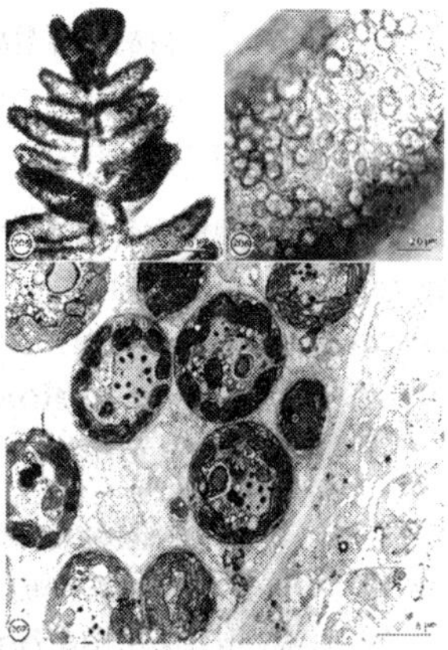

• **zwischenkörper**
a lens-shaped body below the aperture of certain unacetolysed pollen grains.

• **zygomorphic**
bilaterally symmetrical, applied to a flower that has one or more unequal parts.

• **zygomorphic**
(Gr. *zygon*, yoke; *morphe*, shape) said of the corolla or ca-

lyx when divisible into equal halves in one plane only bilaterally symmetrical, with only one plane of symmetry.

- **zygote**

1. a fertilised reproductive cell formed by the merging of two gametes, before any cell division occurs.
2. the product of fertilisation.
3. a reproductive cell formed by the merging of two gametes.

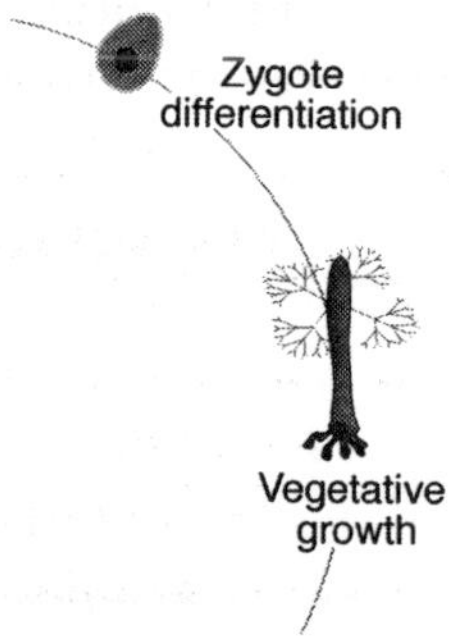

- **zygotic meiosis**

meiosis that occurs during zygote maturation or germination.

Notes

Notes

Notes